Ancient Light

ANCIENT LIGHT

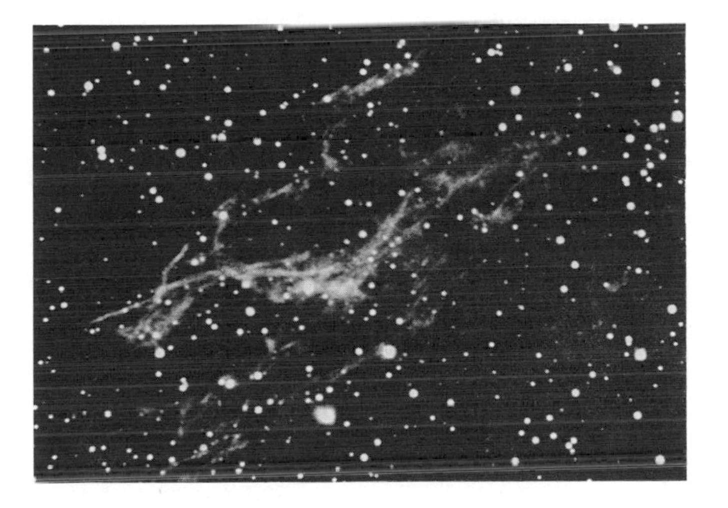

Our Changing View of the Universe

ALAN LIGHTMAN

Harvard University Press
Cambridge, Massachusetts
London, England · 1991

Printed in the United States of America
10 9 8 7 6 5 4 3 2 1

This book is adapted from
Origins: The Lives and Worlds of Modern Cosmologists
by Alan Lightman and Roberta Brawer
published by Harvard University Press

This book is printed on acid-free paper, and its binding materials
have been chosen for strength and durability.

Library of Congress cataloging information is on last page of book.

In memory of
Rabbi James Wax of Memphis
who always thought about
the big picture

Contents

Illustrations

Ancient Light

CHAPTER • ONE

Cosmic Questions

COSMIC questions start inside. Sometime in childhood we become conscious of ourselves as separate from our surroundings; we become aware of our bodies, our own thoughts. Then, we question. How far back can we remember our parents, or anything at all? What was it like before we were born? What would it be like to be dead? We see pictures of our great grandparents and try to imagine their talking and living. In a game, we try to imagine *their* parents, and so on, backward in time, back through the generations, until we don't quite believe that this all could have happened. Yet we are here.

In elementary school we are told that the earth is not flat as it seems, but bends on itself in a huge, mottled ball. We are told that the sun—that bright, small light that circles the sky—is actually far larger than the earth. We are told that the tiny points of light in the heavens are suns themselves. We close our eyes and mentally glide to a star, through the blackness, to stare back at the speck that is earth. In space, events happen slowly, if at all. The sun looks the same, day after day. The stars never vary. In the vast reaches of space, time seems to stretch forward and backward without end, engulfing us, our great grandparents, all human beings, the entire earth. Or perhaps, there is some limit, some enormous boundary that holds time and space. Has the universe existed forever? If not, when and how did it begin? Will it end? Is the universe changing in time? Does space extend infinitely in all directions? How did the matter in the universe come into being?

Every culture has asked these same questions and answered them in some way. Every culture has had its cosmology, its story of how the universe came into being and where it is going.

In recent years astronomers and physicists have addressed the questions of cosmology. Astoundingly, cosmology has become a science. Even so, cosmology is a speculative science. The most widely held cosmological theory, the big bang model, rests on four observational facts: the flight of galaxies, speeding away from one another—discovered in 1929 and interpreted as evidence for the expansion and explosive birth of the cosmos; the approximate agreement between the age of the universe, as gauged by the rate galaxies move away from one another, and the age of the earth, as measured by the radioactive disintegration of uranium ore; the bath of radio waves from space, predicted as a necessary remnant of a hot, younger universe and discovered in 1965; and the overall chemical makeup of the universe—approximately 25 percent helium and 75 percent hydrogen—which can be explained in terms of atomic processes in the infant universe. Aside from these few critical observations, it is theory, assumption, and inference that support the big bang model. Cosmology, of all sciences, requires the most extreme extrapolations in space and in time.

Cosmology today is in a time of upheaval. Recent observations of the locations and motions of galaxies have revealed a universe far more lumpy than previously thought, with galaxies clustering together over vast regions of space. By contrast, the big bang model assumes a uniform universe, whose matter is spread evenly throughout all of space. Cosmologists are not sure how to account for the observed nonuniformity of matter or whether it can be reconciled with the big bang model. In addition, cosmologists have recently come to realize that at least 90 percent of the mass in the cosmos emits no light. This invisible mass, called dark matter, can be detected by its gravitational effects, but its identity is completely unknown. Finally, scientists have developed new theories that allow them to study what the universe may have been like during the first trillionth of a second of its life, a period previously thought to be beyond calculation. The new theories may not be right, but, for the first time, the very birth of the universe has been placed on a scientific footing. Questions considered to lie outside science only twenty years ago are now being pondered by the brilliant scientific minds of the world.

We will begin in Chapter 2 with a brief discussion of early cosmologies. Throughout history each culture's cosmology has reflected its view of the world, and we will see pieces of those earlier world views in modern cosmology. Although based in the methods of science, modern cosmology asks primeval questions and thus sits near the boundary between science, philosophy, and religion. In Chapter 3, we will take up the history of modern cosmology, which begins around 1920, and then move to contemporary observations and theories and the crises of cosmology today.

CHAPTER • TWO

Early Cosmologies

According to the *Enuma Elish,* the ancient Babylonian story of creation, the world began in liquid chaos. There was no sky or ground or even swampy bog. There was only Apsu, the sweet waters, and Ti'amat, the salt waters. In time, the slow seep of silt made Lahmu and Lahamu, who stretched into a giant ring to form the horizon. The upper side of this ring was the god Anshar, and the lower side was Kishar. From Anshar grew Anu, the heaven. From Kishar grew Nudimmut, the earth. But heaven and earth were trapped within the body of Ti'amat, who wished to sleep in silence. Then Marduk, a restless god descended from Kishar and Anshar, was persuaded to do battle with Ti'amat. Ti'amat opened her mouth to swallow him, but Marduk drove in the evil wind to fill her belly, shot an arrow through her heart, and killed her. Then, with his ax, Marduk cut Ti'amat's body in two and separated heaven and earth. Thus, the world was organized.

Carved in the Akkadian language and dating back to at least 2000 B.C., the *Enuma Elish* is one of the oldest recorded cosmologies. It illustrates the human-like characters of early cosmological myths, and its images come from the watery world that the Babylonians knew. Mesopotamia is a country built by silt, at the juncture of the sweet waters of the Tigris and Euphrates, which in turn flow into the salty Persian Gulf. The *Enuma Elish* conveys the human desire to fathom the world, to make a compelling story of how things came to be.

Assyrian cylinder, from 700 B.C., showing worshipper between two gods. The god on the left, holding the ax, is thought to be Marduk.

Buddhist and Hindu legends contain many different versions of creation. One version begins with the god Vishnu, floating on the back of the snake Ananta in the primeval waters. From Vishnu's navel grows a lotus, containing the god Brahma. Upon his birth Brahma recites the scriptures, the four Vedas, from his four mouths, and then he creates the Egg of Brahma. From the Egg comes the universe. The universe is constructed in three layers: Earth, a flattened disk; the atmosphere, associated with wind and rain; and the heavens, the place of sun and fire. Endless cycles of birth, death, and rebirth run through Buddhist and Hindu cosmology. Just as individual people are born, live, die, and are reborn in a different body, so too the universe. At the end of every 4,320,000,000 years, a single day in the life of Brahma, all matter in the universe is absorbed into the universal spirit while Brahma sleeps. During the night of Brahma, matter exists only as potentiality. At dawn, Brahma awakens from the lotus and matter reappears. After 100 of Brahma's years, all is destroyed, including

Brahma himself. After another Brahman century, Brahma is reborn and the entire cycle repeats.

In the West, cosmological speculations gradually shifted from gods and myths to physical mechanisms. Logic and physical reasoning appear in the earliest known Greek cosmological thought, that of Anaximander in the sixth century B.C. According to Anaximander, the stars were compressed portions of air, and the sun was shaped like a chariot wheel, 28 times the size of the earth. The rim of this solar chariot wheel was filled with fire, which escaped through an orifice. When the orifice was plugged, an eclipse occurred. The moon was a circle 19 times the size of the earth, and it too was shaped like a chariot wheel. Aximander's universe was filled with an infinite and ageless substance. Planets and worlds came into being when they broke off from this substance; later, they perished and were reabsorbed by it. In the origin of our own world, a whirpool-like motion caused the heavy materials to sink to the center, where they formed the flattened disk that is earth, while masses of fire surrounded by air were flung to the perimeter, where they formed the sun and the stars. Although individual worlds came and went, the cosmos as a whole was eternal, without beginning or end. It was infinite in time. It was also infinite in space.

Many of Anaximander's ideas can be found in the atomistic theory of Democritus, who lived around 460–370 B.C. In Democritus's cosmology all matter was made out of indestructible microscopic bodies called atoms (from the Greek word *atomos,* meaning indivisible). Atoms differed in properties—for example, some were hard and some were soft, some were smooth and some were thorny—and these differences explained the variation in substances throughout the universe. The Greek atomistic theory provided an explanation for everything, from the nature of wind, to why fish have scales, to why light but not rain passes through a horn, to why corpses smell bad and saffron good. Although substances could change by changing their atoms, the atoms themselves could be neither created nor destroyed. Atoms were eternal. Democritus's atoms were Anaximander's infinite substance.

The atomistic world view had two great assets, clearly stated and applauded by Lucretius in his classic poem *On the Nature of Things* (about 60 B.C.). First, since "nothing can be created from nothing," and "nothing is destroyed into nothing," events cannot happen without physical cause. Human beings need not fear the capricious meddling of the gods. Second, people need not dread everlasting punishment after death. The soul, being composed of atoms like everything else, disperses like the wind after death. No identity remains to be tormented.

When applied to the cosmos at large, the atomistic theory leads to a universe without design and without purpose. Atoms fly blindly through space in every direction. When, by chance, the random paths of a large group of atoms intersect, then a planet or star comes into being. A world thus formed will live for a time, until it disintegrates and returns its atoms to their wanderings. All objects, including people and planets, are just temporary and accidental islands of order in a disorderly cosmos. Our own planet is the same, and it occupies no special place in the universe. Like Anaximander's cosmos, the atomistic universe has no limit in space or in time. A universe made up of indestructible atoms cannot be created or destroyed.

Aristotle's cosmology (ca. 350 B.C.) differed from the atomistic world view in several respects. Aristotle constructed the world out of five elements: earth, water, air, fire, and ether. Nothing was random or accidental. Everything had its natural place and design. The natural place of earth was at the center of the universe, and all earthlike particles in the cosmos drifted to that location. Ether was a divine and indestructible substance; its natural place was in the heavens, where it made up the stars and the other heavenly bodies. Water, air, and fire had intermediate locations. The sun, planets, and stars were attached to rigid spheres, which revolved in perfect circles about the motionless earth. Such revolutions caused night and day. The outermost sphere, the *primum mobile,* was spun by the love of God, while the inner spheres rotated, in sympathy, for the love of God. Thus, unlike the earlier atomistic theory, Aristotle's cosmos had design, and it was bounded in space, extending only

"Expulsion of Adam and Eve from Paradise," by Giovanni di Paolo, depicting the medieval and Aristotelian world view. The earth lies at the center of the system of concentric spheres.

to the outermost sphere. In one important respect the two theories agreed. The universe was eternal. The ether, making up the divine and heavenly bodies, "is eternal, suffers neither growth or diminution, but is ageless, unalterable and impassive." Aristotle's universe was not only eternal; it was also static. This belief in an unchanging cosmos held a firm grip on Western thinking well into the twentieth century.

It was the Polish astronomer Nicolas Copernicus, in 1543, who finally abolished the notion of an earth-centered cosmos. Copernicus demoted the earth to a mere planet orbiting the sun. This

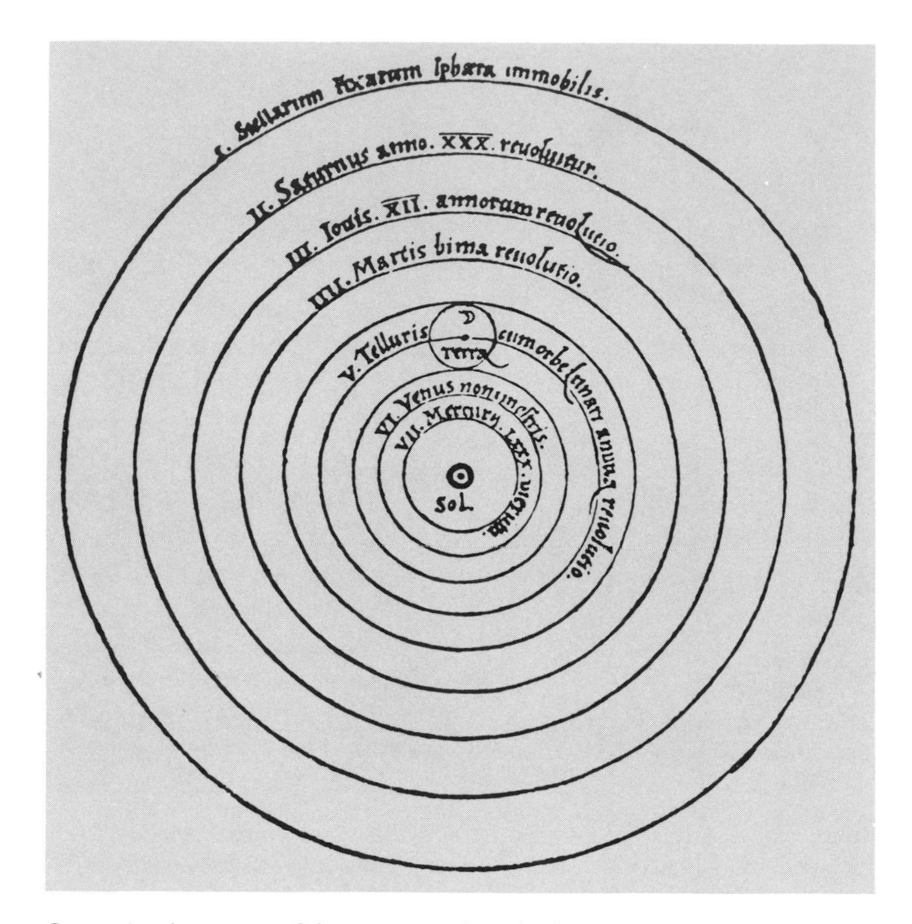

Copernicus's system of the universe, from his book *On the Revolutions of the Heavenly Spheres* (1543). Sol, the sun, lies at the center; terra, the earth, lies in the third orbit about the sun; and the outermost sphere is labeled "stellarum fixarum sphaera immobilis," the immobile sphere of the fixed stars.

important change gave a much simpler explanation to the observed motions of the planets, at the cost of rejecting the intuitive feeling that the earth is at rest. However, Copernicus could not let go of many of the venerable features of the Aristotelian view. Planetary orbits were still composed of perfect circles, as befitted heavenly bodies. And although the earth had been removed from its central location, our sun took its place near the center of the universe. The universe was still made for human beings. As the great German astronomer Johannes Kepler remarked a half century later, our own sun is the most luminous star in the cosmos, "for if there are globes in the heaven similar to our Earth, do we vie with them over who occupies the better portion of the universe? For if their globes are nobler, we are not the noblest of rational creatures. Then how can all things be for man's sake? How can we be the master of God's handiwork?" Copernicus's universe was still spatially bounded by an outermost shell of stars. Like Aristotle, Copernicus continued to believe that the stars were fixed and unchanging. He explained his view in this way: "The state of immobility is regarded as more noble and godlike than that of change and instability, which for that reason should belong to the Earth rather than to the universe." Like Aristotle, Copernicus believed that one set of laws applied to terrestrial phenomena, while another to the "godlike" heavenly bodies.

A disciple of Copernicus, the British astronomer Thomas Digges, succeeded in prying the stars lose from their crystalline spheres and scattering them through infinite space. Set forth in "A Perfit Description of the Caelestiall Orbes" (1576), this notion had an enormously liberating effect on cosmological thought. Now stars could be physical objects. Now stars would be subject to the same physical laws we observe here on earth.

The universality of physical law was given its crowning expression by Isaac Newton. In his *Principia* (1687), Newton applies his new law of gravity equally to the arcs of cannon balls, the orbits of moons and planets, and the trajectories of comets, computing their expected paths in elaborate detail. The master logician was also highly religious. In the very same *Principia,* Newton equates space to the body of God: "The Supreme God is a Being eternal, infinite, absolutely perfect . . . He endures forever and is everywhere present; and by existing always and everywhere, he constitutes duration and space." Newton also argues that "this most beautiful system of sun, planets, and comets could only proceed from the counsel and dominion of an intelligent and powerful Being." Thus, Newton's universe was one of conscious design. Newton's universe was

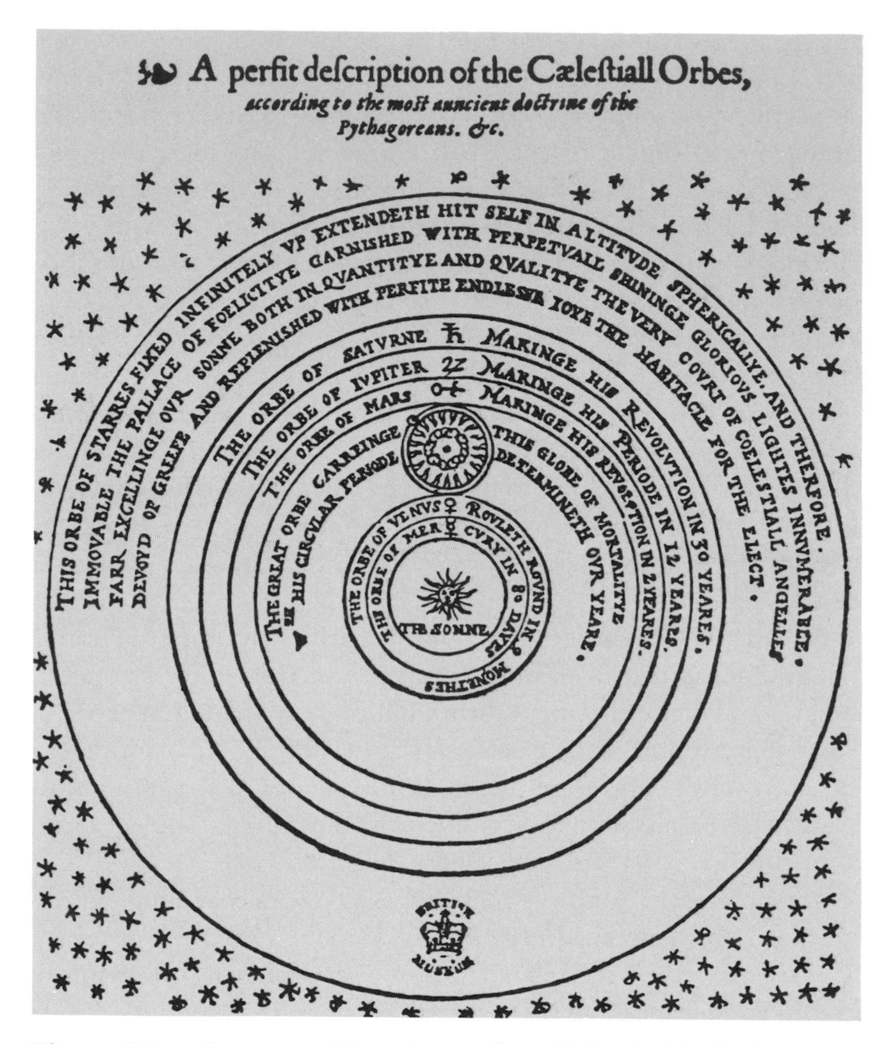

Thomas Digges's system of the universe, from his book *A Perfect Description of the Celestial Orbes* (1576). The stars are scattered throughout space, beyond the outermost orbit of the planets.

also static, on average. In a letter to the theologian Richard Bentley in 1692, Newton argued that the universe could not be globally expanding or contracting because such a motion would necessarily require a center, just as an explosion has a center. However, matter scattered throughout an *infinite* space does not define any center. Therefore, working backwards, the cosmos as a whole must be static. Whether Newton himself was more persuaded by this logical argument or by his religious beliefs, he ended up supporting the Aristotelian tradition of a cosmos without change. That tradition, unchallenged even by Einstein, was not questioned until the late 1920s.

C H A P T E R • T H R E E

The Birth of Modern Cosmology

MODERN theories of cosmology date back to 1917, when Albert Einstein published a pioneering theoretical paper. Using his new theory of gravity, called general relativity, Einstein proposed the first detailed model for the large-scale structure of the universe. Between its publication in 1915 and 1917, the theory of general relativity had been tested by only a single observation, the orbit of the planet Mercury. Einstein's new theory of gravity passed this test with flying colors, explaining a tiny effect in the orbit that could not be accounted for by Newton's older theory. The application of general relativity beyond the solar system, however, remained uncertain. Although Einstein understood that gravity was the dominant force for describing the cosmos at large, he had little working knowledge of astronomy. Not a single astronomical number appears in Einstein's paper on cosmology.

Einstein made two critical assumptions: the universe does not change in time, and the matter of the universe is evenly scattered through space. Given these two assumptions and his mathematical theory of gravity, Einstein was able to derive equations describing the overall structure of the universe.

There was no compelling evidence for either of Einstein's starting assumptions. Although astronomical observations were *consistent* with a static universe, many astronomers of the day were aware

Kosmologische Betrachtungen zur allgemeinen Relativitätstheorie.

Von A. Einstein.

Es ist wohlbekannt, daß die Poissonsche Differentialgleichung

$$\Delta\phi = 4\pi K\rho \qquad (1)$$

in Verbindung mit der Bewegungsgleichung des materiellen Punktes die Newtonsche Fernwirkungstheorie noch nicht vollständig ersetzt. Es muß noch die Bedingung hinzutreten, daß im räumlich Unendlichen das Potential ϕ einem festen Grenzwerte zustrebt. Analog verhält es sich bei der Gravitationstheorie der allgemeinen Relativität; auch hier müssen zu den Differentialgleichungen Grenzbedingungen hinzutreten für das räumlich Unendliche, falls man die Welt wirklich als räumlich unendlich ausgedehnt anzusehen hat.

Bei der Behandlung des Planetenproblems habe ich diese Grenzbedingungen in Gestalt folgender Annahme gewählt: Es ist möglich, ein Bezugssystem so zu wählen, daß sämtliche Gravitationspotentiale $g_{\mu\nu}$ im räumlich Unendlichen konstant werden. Es ist aber a priori durchaus nicht evident, daß man dieselben Grenzbedingungen ansetzen darf, wenn man größere Partien der Körperwelt ins Auge fassen will. Im folgenden sollen die Überlegungen angegeben werden, welche ich bisher über diese prinzipiell wichtige Frage angestellt habe.

§ 1. Die Newtonsche Theorie.

Es ist wohlbekannt, daß die Newtonsche Grenzbedingung des konstanten Limes für ϕ im räumlich Unendlichen zu der Auffassung hinführt, daß die Dichte der Materie im Unendlichen zu null wird. Wir denken uns nämlich, es lasse sich ein Ort im Weltraum finden, um den herum das Gravitationsfeld der Materie, im großen betrachtet, Kugelsymmetrie besitzt (Mittelpunkt). Dann folgt aus der Poissonschen Gleichung, daß die mittlere Dichte ρ rascher als $\frac{1}{r^2}$ mit wachsender Entfernung r vom Mittelpunkt zu null herabsinken muß, damit ϕ im

The title page of Einstein's landmark paper on cosmology, from the journal *Sitzungsberichte der Preussiche Akademie der Wissenschaften* (1917). An English translation of the title is "Cosmological Considerations on the General Theory of Relativity."

that the view seen through big telescopes was only a snapshot, revealing little about the long-term evolution of the cosmos. The observations had nothing to say on this point. On the other hand, the notion of a static universe was deeply ingrained in Western thinking, dating back to Aristotle, and was one of the few astronomical beliefs not overthrown by the Copernican revolution. Einstein's second assumption, of homogeneity, greatly simplified the equations, but it too was made on faith. In fact, as far as astronomers could tell, it was clear that the universe was highly lumpy, with most visible stars gathered up in a great disk called the Milky Way. Until 1918 astronomers had only a poor estimate for the size of the Milky Way; until 1924 astronomers were not sure whether other gatherings of stars, other galaxies, existed in space beyond the Milky Way. Einstein simply assumed that space would appear smooth when averaged over a sufficiently large volume, just as a beach appears smooth when looked at from a distance of a few feet or more, even though it appears grainy when looked at from close range.

Even today, the assumption of homogeneity may be required to manage the mathematics of cosmology. Theorists have succeeded in solving the equations of cosmology only for homogeneous models, except for special and implausible cases. Of course, simple equations and reality are two different things. Nature may not have been so accommodating as to avoid inhomogeneities just because physicists cannot conquer the associated math.

A brief digression on models in science is warranted here. We will encounter a number of cosmological models: the big bang model, the steady state model, the inflationary universe model. A scientific model begins with a real physical object or system, replaces the original object with a simpler object, and then represents the simplified object with equations describing its behavior. Like a toy boat, a scientific model is a scaled-down version of a physical system, missing some parts of the original. Deciding what parts should be left out requires judgment and skill. The omission of essential features makes the model worthless. On the other hand, if nothing is left out, no simplification has been made and the

ALBERT EINSTEIN

Albert Einstein (1879–1955) was born in Ulm, Germany. His father ran a small electrochemical plant. Although he disliked the regimentation of school, Einstein learned much from the mathematics and science books he read on his own. He finished high school in Aarau, Switzerland, and then studied physics and mathematics at the Polytechnic in Zurich. Unable to get an academic job, Einstein was taken on as an examiner at the Swiss Patent Office in Berne in 1902. During the 7 years he spent at his job, Einstein laid the foundations for much of twentieth-century physics, publishing monumental papers in statistical mechanics, quantum mechanics, and special relativity. He received his doctorate from the University of Zurich in 1905. Within a few years, Einstein was renowned and had many offers for professorships. After positions at the German University in Prague and at the Polytechnic in Zurich, he became director of the Kaiser Wilhelm Institute for Physics in Berlin in 1914. Here, he published his work on the theory of general relativity and, in 1917, his pioneering paper on cosmology. After world War I Einstein came under anti-semitic attacks, which grew worse until his departure in 1932 for the Institute for Advanced Study in Princeton, where he remained for the rest of his life. Throughout his life Einstein had deep convictions about freedom and humanity, but he was detached from the day-to-day world. In a speech in 1918 Einstein said, "I believe with Schopenhauer that one of the strongest motives that leads men to art and science is escape from everyday life with its painful crudity and hopeless dreariness, from the fetters of one's own ever shifting desires." In his scientific philosophy, Einstein had a profound belief in the beauty of nature and in the ability of the human mind to discover the truths of nature. However, this discovery could not originate with experiments, but rather as the "free invention of the human mind," after which the mental invention would be tested against experiment and judged accordingly.

analysis is often too complicated to work out. In making a model of a swinging pendulum, for example, we might at first try to include the detailed shape of the weight at the end, the density and pressure of the air in the room, and so on. Finding such a description too complex to manage, we could replace the weight by a round ball and neglect the air completely. This much simpler system, in fact, behaves almost exactly like the original. If, instead, we left out gravity, the resulting theoretical pendulum would not swing back and forth. By solving the equations of a model, predictions can be made about the original physical system and then tested.

In 1922 a Russian mathematician and meteorologist, Alexander Friedmann, proposed cosmological models for a *changing* universe. Friedmann adopted Einstein's theory of gravity and his assumption of homogeneity but rejected his assumptions of stasis, pointing out that it was unverified and nonessential. Beginning with the equations of general relativity, as Einstein had, Friedmann found an alternative solution, corresponding to a universe that began in a state of extremely high density and then expanded in time, thinning out as it did so. Friedmann's model, later rediscovered in 1927 by the Belgian priest and physicist Georges Lemaître, eventually came to be called the big bang model. Einstein reluctantly acknowledged the mathematical validity of Friedmann's evolving cosmological model but initially doubted that it had any bearing on the real universe. In any case, both Einstein's and Friedmann's models were all pencil and paper. Little was known from observations about the true structure or evolution of the universe.

A major stumbling block in all of astronomy, and particularly in cosmology, was the problem of measuring the distances to the stars. When we look at the sky at night, we can perceive width and length, but not depth. From our vantage, stars are just white dots on a black canvas. Some are certainly closer than others, but which ones? Because stars come in a range of luminosities, akin to different wattages of light bulbs, a star of a given *observed* brightness could be either very close and intrinsically dim or very far away and intrinsically bright.

ALEXANDER FRIEDMANN

Alexander Aleksandrovich Friedmann (1888–1925), the son of a musical composer, was born in St. Petersburg, Russia. Studying mathematics and physics at St. Petersburg University, Friedmann graduated in 1910 with a gold medal for his unpublished papers in mathematics. He was to receive his doctorate in 1922. In 1913 Friedmann began his work in meteorology, the main scientific interest of his life, and developed new methods for the theoretical analysis of movements of air masses in the atmosphere. In World War I he worked in aviation and later became director of Russia's first factory for aviation instruments. After 1920 Friedmann worked at the physics observatory of the Academy of Sciences in Petrograd (Leningrad). He was director of researches in the department of theoretical meteorology of the Main Geophysical Laboratory. In 1922 Friedmann published his landmark paper in cosmology, presenting the first theoretical basis for an expanding universe. Friedmann considered that the theory of relativity was essential to an education in physics. The courses he taught at colleges in Petrograd were known for their originality, and his textbooks include *Experiments in the Hydrodynamics of Compressible Liquids* and *The World as Space and Time*. Friedmann died of typhoid fever at the age of 37.

GEORGES LEMAÎTRE

Georges Lemaître (1894–1966) was born in Charleroi, Belgium, and received his doctorate in science and mathematics from the University of Louvain, Belgium, in 1920. He then pursued ecclesiastical studies at the Seminary of Malines, leading up to his ordination in 1923. While at the seminary, Lemaître miraculously found time to produce an unsupervised second thesis on Einstein's new theory of gravity. In the academic year 1924–1925 Lemaître worked as a postdoctoral fellow at the Harvard College Observatory. At a meeting in Washington that year, he heard of Hubble's discovery that the Andromeda nebula was outside our galaxy. As it was already known that most nebulae were speeding away from us, Lemaître interpreted Hubble's result as evidence for a universe in motion. Lemaître hurried back to Belgium and found a new solution to Einstein's equations describing an expanding universe. In his epochal theoretical paper of 1927, Lemaître predicted that the retreating velocity of each galaxy should be proportional to its distance from us—a simple result not pointed out in Alexander Friedmann's earlier and still unknown paper on an expanding universe. For this work Lemaître is sometimes considered the father of the big bang model. In 1931 Lemaître proposed that the entire universe began as a single, giant atom, the "primeval atom," whose gradual disintegrations into smaller and smaller pieces formed nebulae, stars, and finally cosmic rays. Lemaître, always jovial and robust, was dearly loved by his students, who called him the little prince.

Measurements of astronomical distances were placed on much firmer ground around 1912, when Henrietta Leavitt of the Harvard College Observatory discovered a remarkable result for certain stars, called Cepheid variables. It had been known that such stars oscillate in brightness, growing dim, then bright, then dim again, in regular cycles. Leavitt analyzed a group of Cepheids that were clustered about each other and thus *known to be at the same distance.* Within such a cluster, a star that *appeared* twice as bright as another was indeed twice as luminous. Leavitt found that the time for a Cepheid to cycle closely depended on its luminosity. For example, Cepheids 1,000 times as luminous as our sun complete a light cycle every 3 days. Cepheids 10,000 times as luminous cycle every 30 days. Once this behavior has been calibrated for nearby Cepheid stars of known distance and luminosity, it can be used to measure the distance to remote Cepheid stars. By measuring the cycle time of a particular Cepheid star, one can infer its luminosity. By then comparing the star's luminosity to its observed brightness, one can determine its distance, just as the distance to a light bulb may be inferred from its wattage and observed brightness. With Leavitt's discovery, Cepheid stars became distance posts in space.

Cepheid stars found at various locations allowed astronomers in 1918 to measure the size of the Milky Way. In 1924 the American astronomer Edwin Hubble found a Cepheid star in the faint patch of stars known as the Andromeda nebulae, allowing him to measure its distance. He discovered that the Andromeda nebula was a congregation of stars far beyond the Milky Way. Andromeda was a separate congregation of stars, a separate galaxy. Hubble thus became the father of extragalactic astronomy. In the following years, Hubble and other astronomers measured the distances to many of the faint misty patches, called nebulae, that had been seen and puzzled over for centuries. Many were found to be separate galaxies of stars. With these discoveries, galaxies, not stars, became the basic units of matter in the universe.

In cosmology, the first thing that overwhelms us is the vastness of space. To appreciate cosmic distances, we might start close to home. The circumference of the earth is about 24,000 miles, the

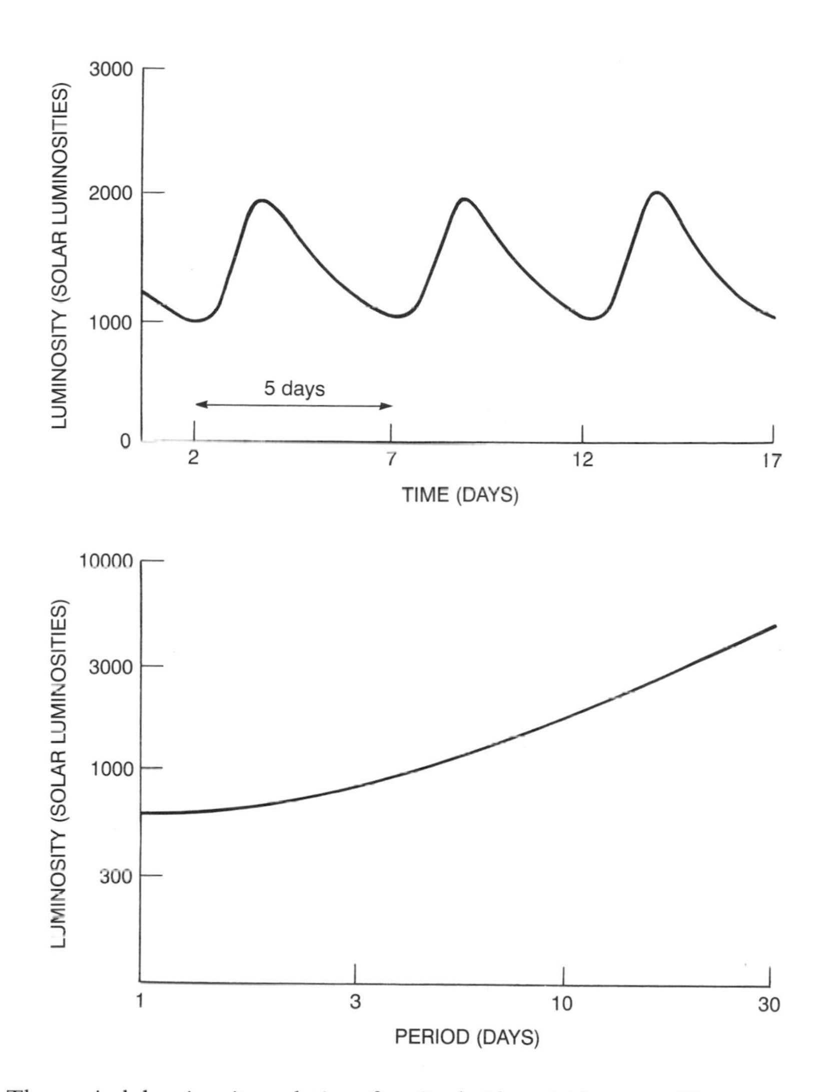

The period–luminosity relation for Cepheid variable stars. The top curve shows the variation in time of the luminosity of a typical Cepheid. The luminosity of this particular star varies between about 1,000 and 2,000 times that of the sun, on a cycle of about 5 days. The bottom curve shows how the *average* luminosity of Cepheid stars varies with their cycle time, or period. The particular Cepheid star shown in the top curve corresponds to one point in the bottom curve.

HENRIETTA LEAVITT

Henrietta Swan Leavitt (1868–1921) was born in Lancaster, Massachusetts, and was one of seven children of a prominent Congregationalist minister. She studied astronomy at what is now Radcliffe College and received her A.B. degree there in 1892. A few years later she joined the Harvard College Observatory, where she worked for the rest of her life. Following a program suggested by Edward Pickering, director of the Observatory, Leavitt became one of the first astronomers to systematically compare the brightness of stars using photographic plates, which have greater sensitivity to blue light than does the human eye. Leavitt was particularly interested in variable stars. Such stars change in brightness over a period of days to months. Over her career she discovered 2,400 variable stars. Her most important discovery was the period-luminosity relation for Cepheid variables, used by many subsequent astronomers to measure cosmic distances. Leavitt was one of several women, including Williamina Fleming and Annie Jump Cannon, who worked at the Harvard College Observatory in the early part of the twentieth century. Like Cannon, Leavitt was deaf. Her work on variable stars was vitally important but without glamour, requiring extremely meticulous and untiring study of hundreds of photographic plates to find the brightening and dimming of the tiny spots that were stars. Throughout her life, Leavitt was dedicated and religious and held to the strict values of her Puritan ancestors.

The Andromeda galaxy, also known as M31.

distance to the moon about 250,000 miles, the distance to the sun about 100 million miles. The distance to the star nearest the sun, Alpha Centauri, is about 25 trillion (25,000,000,000,000) miles. All of these distances were well estimated by the eighteenth century. To measure further distances, it is convenient to use the light year, which is the distance that light travels in a year, about 6 trillion miles. In these terms, Alpha Centauri is about 4 light years away. Our galaxy, the Milky Way, is about 100,000 light years in diameter. In other words, it takes a light beam 100,000 years to cross from one side of the Milky Way to the other. The nearest galaxy to us, Andromeda, is about 2 million light years away.

A typical galaxy, like our Milky Way, contains about 100 billion stars, which orbit one another under their mutual gravity. Galaxies come in a variety of shapes. Some are nearly spherical, while others, like the Milky Way, are flattened disks with a bulge in the middle. It takes our sun about 200 million years to complete one orbit about the center of the Milky Way. On average, galaxies are separated by

The Sombrero galaxy, also known as M104.

about 10 million light years, or about 100 times the diameter of one galaxy. Galaxies, therefore, are isolated islands of stars, surrounded by mostly empty space. Einstein's assumption of homogeneity would have to be tested on volumes of space that encompassed many galaxies.

In 1929 Hubble made what was perhaps the most important discovery of modern cosmology: the universe is expanding. Using data taken from a telescope at Mt. Wilson, California, Hubble concluded that the other galaxies are moving outward from us in all directions. Two kinds of measurements are needed in this analysis: the speed and the distance of neighboring galaxies. It had been known since the early 1900s that many of the nebulae were in motion, speeding away from the earth. This result had been determined by a technique known as the Doppler shift. Galaxies, like all sources of light, emit light of particular colors (wavelengths), related to the chemical composition of the galaxy. When a source of light is in motion, its colors shift, analogously to the shift in pitch of a moving source of sound. For example, the pitch of a train's

EDWIN HUBBLE

Edwin Powell Hubble (1889–1953) was born in Marshfield, Missouri. His father was a lawyer and was in the insurance business. In college, at the University of Chicago, Hubble excelled in mathematics, astronomy, and boxing. A boxing promoter offered to train the tall and barrel-chested Hubble to fight Jack Johnson, the world heavyweight champion, but Hubble instead went to Oxford as a Rhodes scholar. Hubble decided to study law there. Upon his return to the United States, in 1913, he opened a law office in Louisville, Kentucky, but soon left to do graduate work in astronomy at the University of Chicago. During World War I, Hubble enlisted as a private and finished as a major. After the war, at the age of 30, Hubble began his famous work at the newly operational 100-inch telescope at Mt. Wilson, California. Hubble's first major discovery was a Cepheid variable star in the Andromeda nebula, establishing that object as the first known galaxy outside the Milky Way. Hubble also created a classification scheme for galaxies based on their shape. In 1929 Hubble made his greatest discovery: the speed-proportional-to-distance relation for receding galaxies, indicating that the universe is expanding. This discovery confirmed the earlier theoretical prediction of Georges Lemaître. After World War II Hubble devoted much of his energies to the upcoming program of the 200-inch telescope at Mt. Palomar. When the telescope was completed in 1949, Hubble was the first to use it, continuing his study of the measurement of cosmic distances. In addition to science, Hubble was interested in art, athletics, and fly fishing, and he regularly fished in the Rocky Mountains. In science, he believed in the uniformity of nature and the universality of physical principles. Hubble carried this philosophy to the limit in his study of galaxies billions of light years away. Of this philosophy, he said it "is the fundamental assumption in all extrapolations beyond the limits of known and observable data, and all speculations which follow its guide are legitimate until they become self-contradictory."

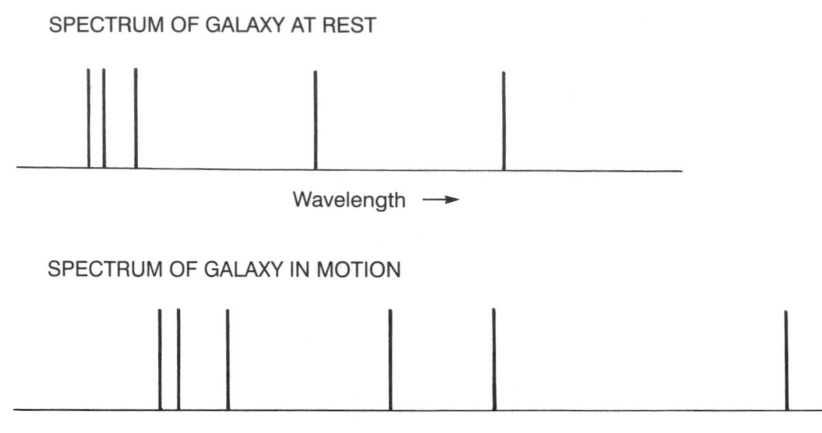

The redshift effect. After the light from a galaxy is passed through a prism, each emitted color appears as a vertical line at a certain wavelength. Motion of the galaxy causes each emitted color to shift its wavelength by a fractional amount proportional to the speed of motion. The fourth line to the right in the bottom figure, for the galaxy in motion, corresponds to a color not emitted by the galaxy at rest. However, the first 3 lines, and their relative placement, can be unambiguously identified with the first 3 lines of the galaxy at rest and thus used to compute the wavelength shift and hence speed of the galaxy in motion.

whistle drops when the train moves away and rises when the train moves closer. In light, the analogue of pitch is color. If a light source is moving closer, its colors are shifted down in wavelength, toward the blue end of the spectrum; if the source is moving away, its colors are shifted up, toward the red. From the amount of the shift, one can infer the speed of the moving source of light. Although the effect in light is usually tiny, sensitive instruments can detect it.

If one assumes that the same basic chemicals are present in all galaxies, then the emitted colors of galaxies at rest should be the same. It had been found by about 1920 that the tell-tale colors of many of the nebulae were shifted to the red, showing that they

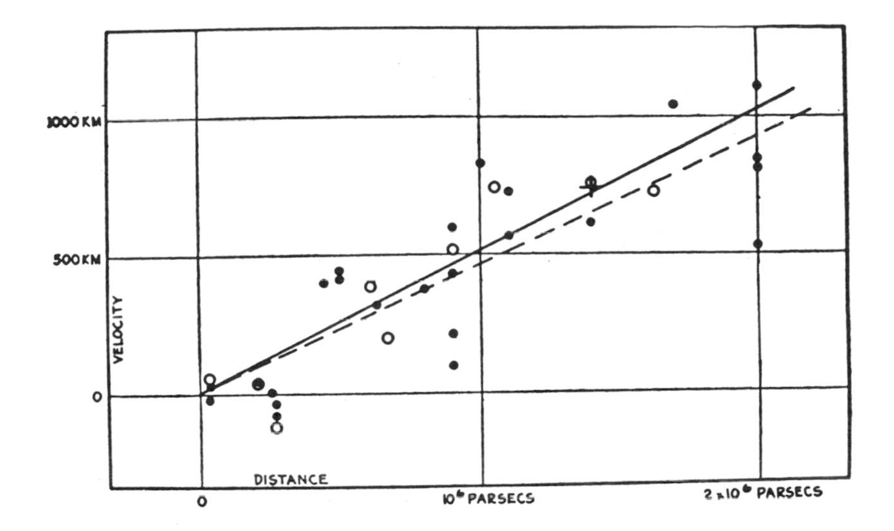

Graph of distance versus velocity for nearby galaxies, taken from Edwin Hubble's landmark paper "A Relation between Distance and Radial Velocity among Extra-Galactic Nebulae" (1929), which appeared in *Proceedings of the National Academy of Sciences*. The vertical axis is the recessional speed in kilometers per second. The horizontal axis is the distance, in parsecs (about 3 light years). Each black dot is an individual galaxy; each open circle is a group of galaxies. The solid and dashed lines are curves fit to the data of the dots and circles. The approximately linear relation between distance and velocity is "Hubble's Law."

were speeding away from us. This change in color of cosmic objects came to be called the redshift.

By using Cepheid stars to measure the distances to about 18 nebulae, Hubble found that the nebulae were entire galaxies, lying beyond the Milky Way. More important, he discovered that the distance to each galaxy was proportional to its recessional speed: a galaxy twice as far from us as another galaxy was moving outward twice as fast.

This last quantitative result was just as had been predicted for a uniformly expanding and homogeneous universe. A simple example, with homemade equipment, shows why. Place equally spaced ink marks on a rubber band, declare one ink mark to be

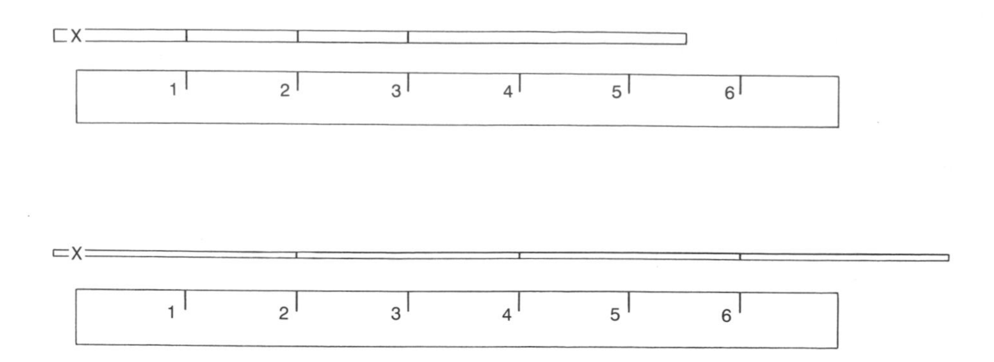

The universe on a rubber band, demonstrating Hubble's law as a consequence of a uniformly stretching medium. Upon stretching the rubber band and keeping the cross at a fixed point as a reference, each ink mark moves a distance proportional to its initial distance from the reference point.

your reference point (for example, the Milky Way), and measure all distances and motions relative to it. Hold the reference mark *fixed* against a ruler, say at the 0 inch mark, and then stretch the two ends of the rubber band. Upon stretching, you will find that each ink mark moves a distance proportional to its initial distance from the reference mark. For example, when the ink mark initially 1 inch away moves to 2 inches away, the ink mark initially 2 inches away moves to 4 inches away. Since these increased distances are accomplished in the same period of time, the second ink mark moves twice as fast as the first. Speed is proportional to distance. In fact, *any* uniformly stretching material produces the law that speed is proportional to distance. If the material is lumpy, so that some places stretch at a faster rate than others, then speed is no longer proportional to distance. Conversely, the proportionality of speed to distance means that the material is uniformly stretching. It is also easily seen that the expansion has no center or privileged position. *Any* ink mark can be chosen as the reference mark, and the result is the same: the other ink marks move away from it with speeds proportional to their distances. No ink mark is special. The view is the same for all.

Replace the ink marks by galaxies and the stretching rubber band by the stretching space of the universe, and you arrive at Hubble's result. Galaxies are moving away from us because space is stretching uniformly in all directions, carrying the galaxies along with it. Hubble's discovery in 1929 gave strong observational support for cosmological models in which the universe is uniformly expanding. The static universe of Einstein was ruled out. The big bang model of Friedmann and Lemaître was supported.

If the galaxies are moving away from each other, then they would have been closer together in the past. At earlier times, the universe was denser. If we assume that this backward extrapolation can be continued, then there was some definite moment in the past when all the matter of the universe was crammed together in a state of almost infinite density. From the rate of expansion, astronomers can estimate when this point in time occurred: about 10 to 15 billion years ago. It is called the beginning of the universe, or the big bang. The original estimates of Hubble, in error owing to various technical problems, gave an estimate of about 2 billion years for the age of the universe. For simplicity, we will assume 10 billion years in all subsequent discussion.

There is a completely independent method for determining the age of the universe. That method involves the earth. Radioactive dating of terrestrial uranium ore, developed about two decades before Hubble's discovery, suggests that the earth is about 4 billion years old. What should this have to do with the age of the universe? Most theories of the formation of stars and planets indicate that our solar system could not be a lot younger than the universe. In astronomy, where the ages of things span many factors of 10, 4 billion years is considered very close to 10 billion years. The match is good. Thus with two totally different methods, one using the outward motions of galaxies and one using rocks underfoot, scientists have derived comparable ages for the universe. This agreement has been a powerful argument in favor of the big bang model.

Cosmology and geology share even more. Digging down to deeper layers of the earth is traveling back in time, back into our human past. Peering out to greater distances in space is also traveling back in time. When our telescopes detect a galaxy 10 million light years away, we see that galaxy as it was 10 million years ago; we see ancient light, which has been traveling 10 million years to get from there to here. When we detect a more distant galaxy, we gaze upon an even older image, we see even older light. Cosmological observation is a kind of excavation, a search for origins above the ground, a glimpse of not an earlier earth but an earlier universe.

C H A P T E R • F O U R

The Big Bang Model

THE BIG bang model logically follows from Einstein's theory of gravity and a small number of assumptions. This model provides a mathematical description for the evolution of the universe. According to the big bang model, the universe began in a sort of explosion, starting from infinite density and temperature, and has been expanding, thinning out, and cooling ever since. The beginning was not like an ordinary explosion, in which debris flies out into a surrounding region of nonmoving space. Instead, the big bang explosion occurred everywhere. There was no surrounding space for the universe to move into, since any such space would be part of the universe. The concept boggles the imagination, but it is a little easier to visualize if one pictures individual particles in the universe. Since the big bang, all particles in the universe have been moving away from one another, carried along by the expansion of space, just as the ink marks move apart on a stretching rubber band. (To be more precise, all *sufficiently distant* particles have been moving away from one another since the big bang. Particles highly clustered and close together are affected by their mutual attractive forces and do not participate in the overall expansion of the universe. For example, the atoms in your body are held in place by electrical forces and do not expand away from one another. Likewise, the stars in a galaxy are held in place by their mutual gravitational attraction and do not expand away from each other.)

Even though the universe expands, its parts tug on one another owing to gravitational attraction, and this slows down the expansion. The competition between the outward motion of expansion and the inward pull of gravity leads to three possibilities for the ultimate fate of the universe. The universe may expand forever, with its outward motion always overwhelming the inward pull of gravity, just as a rock thrown upward with sufficient speed will escape the gravity of the earth and keep traveling forever. Such a universe is called an open universe. A second possibility is that the inward force of gravity is sufficiently strong to halt and reverse the expansion, just as a rock thrown upward with insufficient speed will reach a maximum height and then fall back to earth. For a universe of this type, called a closed universe, the universe reaches a maximum size and then starts collapsing, toward a kind of reverse big bang. Such universes have both a beginning and an end in time. The final possibility, called a flat universe, is midway between a closed and open universe and is analogous to the rock thrown upward with precisely the minimum speed needed to escape from the pull of the earth. Flat universes, like open universes, keep expanding forever.

All three possibilities are allowed by the big bang model. Which one holds for our universe depends upon the manner in which the cosmic expansion began, in the same way that the path of the rock depends on the rock's initial speed relative to the strength of the earth's gravity. For the rock, the critical initial speed is 7 miles per second. Rocks thrown upward with less than this speed will fall back to earth; rocks with greater initial speed will never return. Likewise, the fate of the universe was determined by its initial rate of expansion relative to its gravity. Even without knowledge of these initial conditions, we can infer the fate of our universe by comparing its *current* rate of expansion to its *current* average density. If the density is greater than a critical value, which is determined by the current rate of expansion, then gravity dominates; the universe is closed, fated to collapse at some time in the future. If the density is less than the critical value, the universe is open. If it is precisely equal to the critical value, the universe is flat. The ratio of

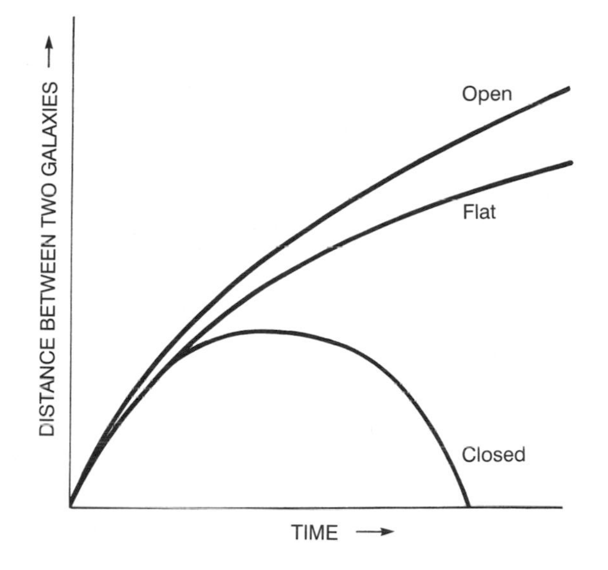

Expansion of the universe in time for closed, open, and flat cosmologies. The expansion may be measured by the distance between any two distant galaxies. In a closed universe, the universe expands at first and then contracts.

the actual density to the critical density is called omega. Thus, the universe is open, flat, or closed depending on whether omega is less than 1, equal to 1, or larger than 1, respectively.

In principle, omega can be measured. The rate of expansion of the universe is estimated by measuring the recessional speed of a distant galaxy (found by its redshift) and dividing by the distance to the galaxy. In a uniformly expanding universe, as we have seen, the outward speed of any galaxy is proportional to its distance; thus, the ratio of velocity to distance is the same for any galaxy. The resulting number, called the Hubble constant, measures the current rate of expansion of the universe. According to the best measurements, the current rate of expansion of the universe is such that it will double its size in approximately 10 billion years. This corresponds to a critical density of matter of about 10^{-29} grams per cubic centimeter, the density achieved by spreading the mass of a poppy seed through a volume the size of the earth. (The notation 10^{-29} is shorthand for a decimal point followed by 28 zeros and a one; 10^{15} stands for a one followed by 15 zeros; and so on.) The best measured value for the actual average density—obtained by telescopically examining a huge volume of space containing many galaxies, estimating the amount of mass in that volume by its gravitational effects, and then dividing by the size of the volume—is about 10^{-30} grams per cubic centimeter, or about one tenth the critical value. This result, as well as other observations, suggests that our universe is of the open variety.

However, there are uncertainties in these estimates, mostly connected with inhomogeneities in the universe and uncertanties in cosmic distances, and omega is difficult to measure in practice. If the universe were precisely homogeneous and uniformly expanding, then the rate of expansion of the universe could be determined by measuring the recessional speed and distance of any galaxy, near or far. Conversely, the distance to any galaxy could be determined from its redshift and the application of Hubble's law. (Roughly speaking, the distance to a galaxy is 10 billion light years multiplied by the fractional increase in wavelength of its detected light.) However, the universe is not completely homogeneous. Because of local inhomogeneities, the rate of expansion of the universe and the average density of matter should be measured over as large a region as possible, after which we must assume that such a region is typical of any large volume of the universe. Accurate distances to galaxies are needed for both of these measurements. The rate of cosmic expansion, for example, is obtained by dividing the recessional speed of a galaxy by its distance, if the latter is known. Individual Cepheid stars cannot be used to measure distances beyond about 30 million light years, because the stars become too dim. Instead, entire galaxies must be used as "standard candles," that is, objects of known luminosity. Unfortunately, galaxies, like stars, come in a wide range of luminosities. There are no standard candles. The best that can be done is to search for some empirical relation between the luminosity of a galaxy and another observed property, such as the orbital speed of its stars. (Such relations are analogous to the relation between the luminosity and light period of a Cepheid star.) After determining and calibrating such a relation for nearby galaxies, where distances can be measured by other means, the method can then be applied to much farther galaxies.

The pitfall is that looking to greater distances in space is equivalent to looking back in time. The light we see today from distant galaxies was emitted when they were much younger and has been traveling for hundreds of millions or billions of years to reach us. By contrast, the light from nearby galaxies has been traveling for a much shorter time and therefore shows them at a much later stage of their evolution. In other words, a distant galaxy *as now observed* may be very different from the more mature galaxies nearby that have been used to calibrate the luminosity-orbital speed relation, and the relation may not apply to the distant galaxy very well. The problem is not unlike trying to extend the relationship between height and weight found for sixty-year-olds to twenty-year-olds. Many modern astronomers have devoted their work to understanding the long-term changes of galaxies. But at present, we have not found any astronomical objects as well understood and reliable as Cepheid stars for using as standard candles.

The difficulty in measuring large distances also limits our ability to determine the average density of the universe. The average density of mass in the largest volumes of space we have measured, extending in size up to a few hundred million light years, is estimated by how the motions of galaxies are affected by local concentrations of mass. If the mass of the universe were spread smoothly, then each galaxy would move directly outward from us, with a speed exactly proportional to distance. Indeed, this is approximately what we see. However, in "local" regions, of 10 or 100 million light years, the cosmic mass is clumped into galaxies and clusters of galaxies. The "lumpy" gravity of such mass clusters bends and alters the course of nearby galaxies—just as, for example, the ball in a pinball machine gets deflected this way and that as it runs into bumpers, even though falling down on the average. A comparison of the "peculiar" motions of the galaxies with the "normal" motions expected for a completely smooth universe, together with a knowledge of the "bumpers" producing the irregular motions, determines the average density of matter in the region. The difficulty is that only the *total* motion of such galaxies can be measured; to know how much is peculiar and how much normal requires knowledge of the rate of expansion of the universe and the *distance* to the galaxies. (Recall that the normal outward speed is the rate of expansion multiplied by the distance.) If the distance is not known well, the normal speed of expansion will not be known well either.

A different method of determining omega involves gauging how the rate of expansion of the universe has been slowing down in time. This is closely related to measuring both the expansion rate and the average density of matter, since the gravity of the latter is the assumed cause of the slow down. In practice, the expansion rate is measured at greater and greater distances, which probe the universe at earlier and earlier times. Unfortunately, these measurements, which were begun by Edwin Hubble in the 1930s and continued by Allan Sandage of the Mt. Wilson Observatories in the 1950s, require either that accurate distances be determined for very distant objects or that a set of standard candles of known luminosity be found. Thus, they too suffer from the difficulties of measuring large distances in the cosmos and the absence of standard candles.

Despite these uncertainties, cosmologists are fairly sure that the value of omega lies between 0.1 and 2.0. Enough matter has been identified so that omega cannot be less than 0.1. On the upper end, an omega larger than 2, together with the current rate of expansion, would translate to an age of the universe less than the age of the earth as determined by radioactive dating.

Einstein's theory of gravity, which underlies the big bang model, makes a theoretical connection between the evolution of the universe and its size. According to the theory, if the universe is closed, then it has a limited size. One might ask what lies outside the boundary of a universe that has only a limited size. The answer is that a closed universe has no boundary. It bends around on itself, in the same way that the surface of a sphere bends around on itself.

Allan Sandage was born in Iowa City in 1926, educated at the University of Illinois and the California Institute of Technology, and is now on the staff of the Mt. Wilson and Las Campanas Observatories. Sandage has undertaken a number of extensive programs to measure the distances to galaxies, the rate of expansion of the universe, and the deceleration of the expansion. As a young graduate student in the early 1950s, Sandage became the assistant to Edwin Hubble.

Begin walking in a straight line and you come back to where you started. You travel around your entire world, covering a finite distance, but you never fall off an edge or meet a boundary. In three dimensions, this picture resists the imagination, but it can be expressed mathematically. Open and flat universes, by contrast, have unlimited size and extend infinitely in all directions. There is a further distinction between closed, flat, and open universes. Flat universes obey Euclidean geometry. For example, the three angles of a triangle formed by connecting three galaxies with straight lines add up to 180 degrees. In closed universes, the angles of such a triangle add up to *more* than 180 degrees; in open universes the sum is *less* than 180 degrees. Closed and open universes have a non-Euclidean geometry, first explored by mathematicians in the nineteenth century.

People are often confused about what it means to speak of the expansion of open or flat universes, which already extend infinitely in space. Expansion means that the distance between any two galaxies is increasing. When we say that the universe is currently expanding at a rate such that it will double in size in 10 billion years, we mean that the distance between any two widely separated galaxies will double in 10 billion years. Such a definition has sense in open, flat, or closed universes.

The overall geometry or size of the universe has not been directly measured. The measured quantities are the rate of expansion and the average density. Only after combining these measured quantities with the big bang theory and its mathematics can we infer the geometry and the fate of the universe. Thus, much depends upon the theory and its key assumptions.

It is also important to realize that even if the universe is of infinite size, only a limited volume, called the observable universe, is *visible* to us at any moment. That is because we can see only as far as light can have traveled since the big bang. As we look farther into space, we are seeing light that has been traveling longer to reach us and therefore that was emitted earlier. When we look at the Andromeda galaxy, for example, we see light that was emitted 2 million years ago; when we look at the Virgo cluster of galaxies, we see light that was emitted 50 million years ago. Eventually, at some distance, the light just now reaching us was emitted at the moment of the big bang. That distance marks the edge of the currently observable universe. We cannot see farther because there hasn't been time for light to travel from there to here since the big bang. Today, the observable universe extends out about 10 billion light years, the distance light can travel in 10 billion years. A billion years from

HORIZONS IN AN EXPANDING UNIVERSE

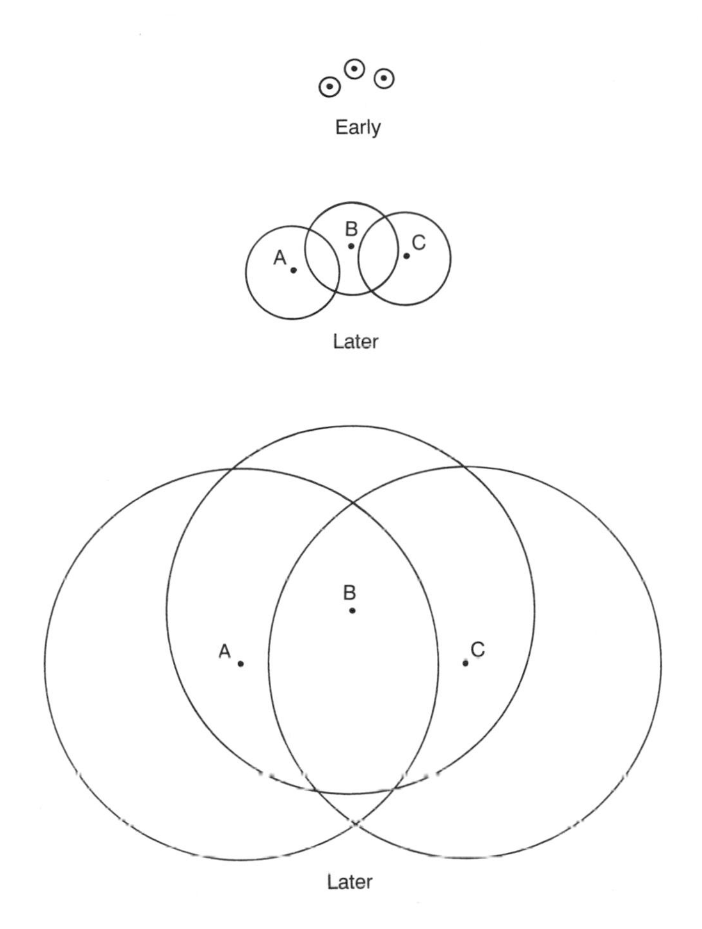

Horizons in an expanding universe. The circle around each dot is the horizon for that dot—the most distant region that it can have communicated with since the big bang. As soon as the horizon of one point encompasses another point, the two points come into communication and can see each other. As the universe expands, various points, such as those labeled A, B, and C, move away from one another, but the horizon of each point expands even faster, so that points initially outside of one another's horizon eventually pass within and into communication. At any moment, the horizon of each dot (location) marks the edge of the observable universe from that location.

now, when the universe is 11 billion years old, the observable universe will extend to 11 billion light years; human beings, if they are still around, will be able to see to a distance of 11 billion light years. New regions of the universe, now beyond our horizon, will have come into view. We can never see further back in time than the big bang, but as time goes on we see more and more of the universe as it was at the big bang. Each day the observable universe grows a little larger. Each day, the light emitted from slightly more distant objects has had the needed time to reach our telescopes.

The big bang model does more than relate the evolution of the universe to its mass density and geometry; the model describes the broad history of the universe. Imagine a movie of cosmic evolution played backward in time, starting from the present. The universe contracts. The galaxies move closer and closer together and turn into aimless blobs of gas. As the universe grows denser and denser, the gas blobs merge. Individual galaxies and even individual stars lose their identity, and the matter of the universe begins to resemble a gas. Like any gas being compressed, the cosmic gas becomes hotter and hotter. Eventually, at a temperature of about ten thousand (10^4) degrees centigrade, the heat becomes so high that atoms cannot retain their electrons, and they disintegrate into atomic nuclei and freely roaming electrons. At a still earlier age, as the big bang gets nearer, the atomic nuclei themselves disintegrate into protons and neutrons under the high heat. At an even earlier time, when the temperature has reached about 10^{13} degrees centigrade, each proton and neutron disintegrates into three elementary particles called quarks. The universe becomes a sea of careening subatomic particles.

The big bang model is quantitative. It specifies the average density, expansion rate, and temperature of the universe at each point in time, given the measured values of those quantities today. According to the theory, the temperature of the universe was about 10 billion (10^{10}) degrees centigrade and its density was about one hundred thousand (10^5) grams per cubic centimeter one second after the big bang. At that moment, the universe consisted of a very hot gas of subatomic particles, uniformly filling space. By the

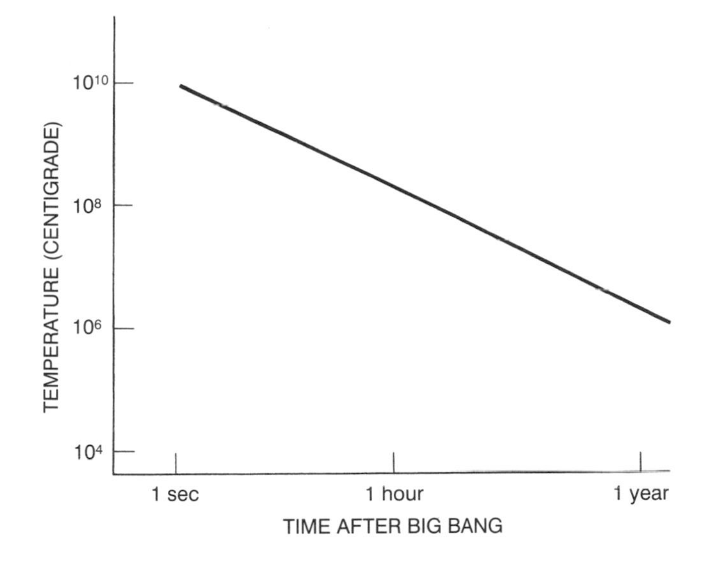

The temperature of the universe during the first year after the big bang, as calculated in the big bang model. (The temperature before the first second is not shown.)

time the universe was about 30 million years old—the epoch thought by some scientists to be roughly when the first galaxies began forming—its temperature and density had dropped to about zero degrees centigrade and 10^{-25} grams per cubic centimeter, respectively. (Absolute zero is −273 degrees centigrade. The measured cosmic temperature today is about −270 degrees centigrade, or 3 degrees above absolute zero, and is still dropping.)

In addition to providing an explanation for the observed expansion and age of the universe, the big bang model has successfully met two other major tests against observations. It explains why the universe is approximately 75 percent hydrogen and 25 percent helium, as observed. (The heavier chemical elements, such as oxygen and carbon, make up only a trace of the total mass in the universe.) The big bang model also predicted that space should be filled with a special kind of radio waves, created when the universe was much younger. Such cosmic radio waves, called the cosmic background radiation, were discovered in 1965, after their prediction. The big bang's successes with helium and with cosmic radio waves—the first a good explanation of a previously known fact and the second a prediction of a to-be-discovered fact—were crucial, not just for the science but for the attitudes of scientists. The agreement between theory and observation on these two phenomena convinced many scientists for the first time that cosmology had some contact with reality, that cosmology was a legitimate science.

According to the big bang model, the universe was once so hot that none of the chemical elements except hydrogen, the lightest element, could exist. Hydrogen is nothing more than a single subatomic particle, a proton. All other elements consist of a fusion of two or more subatomic particles, which could not hold together under the intense heat of the infant universe. As the universe expanded, it cooled. When the universe was a few minutes old, its temperature had dropped to a billion degrees—the critical temperature at which subatomic particles could begin sticking together via the attractive nuclear forces between them. According to theoretical calculations done by Fred Hoyle and Roger Tayler of Cambridge University in 1964 and by Yakov Zel'dovich of the Institute for Cosmic Research in Moscow at about the same time, and then refined by James Peebles at Princeton University in 1966 and by Robert Wagoner, William Fowler, and Fred Hoyle at the California Institute of Technology in 1967, nuclear fusion in the few minutes after the big bang should have converted about 25 percent of the mass of the universe into helium, the next lightest element after hydrogen. In the 1980s, David Schramm and collaborators at the University of Chicago, improving upon the earlier work of Wagoner, Fowler, and Hoyle, accurately calculated the expected amount of lithium produced in the big bang. Lithium is the next lightest chemical element after helium and makes up only about 0.0000001 percent of the observed mass of the universe. It is believed that all heavier elements after lithium were manufactured much later, in the nuclear reactions at the centers of stars. Remarkably, the theoretical accounting of hydrogen, helium, and lithium is in good agreement with the observed abundances of those elements in space.

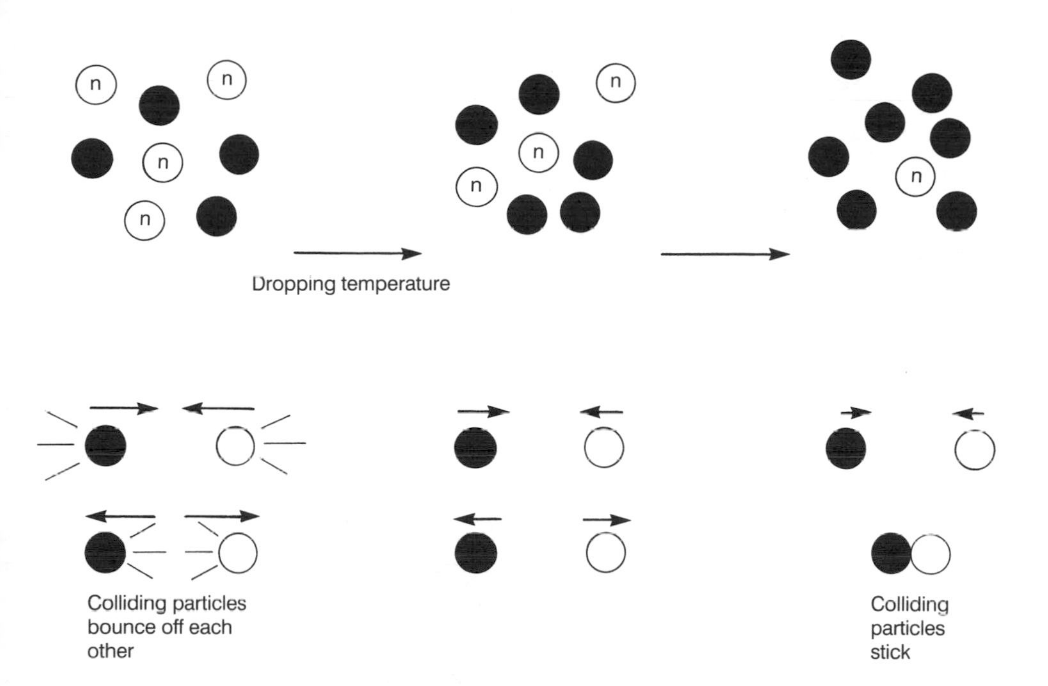

Determination of helium-to-hydrogen ratio in the infant universe. Protons are black circles; neutrons are open circles labeled by n. A hydrogen nucleus has 1 proton; a helium nucleus has 2 protons and 2 neutrons. At early times there were equal numbers of neutrons and protons. When these particles collided, they could not fuse and hold together because the temperature was too high. As time went on, two things happened: the neutrons began turning into protons, gradually decreasing the neutron-to-proton ratio, and the temperature continuously decreased. At a critical moment, when the universe was about a minute old and when the cosmic temperature had dropped to about 1 billion degrees, the neutrons and protons could hold together when they collided, creating helium. All of the available neutrons were then locked up in helium nuclei, with two protons paired with every two neutrons. Left over protons became hydrogen. The resulting fraction of hydrogen to helium would be determined by the ratio of neutrons to protons at the moment just before fusion began.

James Peebles was born in Winnipeg, Manitoba, Canada, in 1935, educated at the University of Manitoba and Princeton University, and is now professor of physics at Princeton. Among other contributions to cosmology, Peebles pioneered calculations of gravitational clustering of matter and has become a leading theoretician of the "gravitational hierarchy" model of structure formation. Peebles was a student of Robert Dicke's.

Robert Dicke was born in St. Louis, Missouri, in 1916, educated at Princeton University and the University of Rochester, and is now professor of science emeritus at Princeton. Dicke's contributions to cosmology include a prediction of the cosmic background radiation, the first use of the anthropic principle, and the first statement of the flatness problem. Dicke is one of a small fraction of physicists who are outstanding at both theoretical and observational work.

Fred Hoyle was born in Bingley, Yorkshire, England, in 1915, educated at Cambridge University, and now lives in Bournemouth England. Hoyle pioneered the theoretical calculations of helium production in the early universe and also developed the steady state model, rivaling the big bang model. An advocate of the steady state model for much of his career, Hoyle coined the phrase "big bang" during a series of radio talks he gave in England in the late 1940s.

The other important experimental confirmation of the big bang theory, the cosmic background radiation, was first predicted by Ralph Alpher, George Gamow, and Robert Herman at George Washington University in 1948, and later independently predicted by Robert Dicke, James Peebles, P. G. Roll, and David Wilkinson of Princeton University in 1965. Both of these groups pointed out that when the universe was a few seconds old and younger, a special kind of radiation, called blackbody radiation, would have been produced throughout space. Such radiation arises in any system of subatomic particles that collide with each other very rapidly, as would have been the case in the high heat of the infant universe. Small amounts of blackbody radiation are also produced today, in isolated regions such as in stars, but the universe now is much too cool as a whole to produce blackbody radiation filling all space. Blackbody radiation is easily identifiable by its universal spectrum of colors, that is, the amount of energy at each wavelength. Such radiation can be uniquely characterized by a single parameter, which corresponds to the temperature of the radiation. According to theoretical calculations, blackbody radiation should have been produced uniformly through space in the early universe and would have continued bouncing off subatomic particles until the universe was about 300,000 years old, when electrons and atomic nuclei combined to make atoms. After that, the radiation would have

Arno Penzias (foreground) and Robert Wilson standing next to the radio antenna that first detected the cosmic background radiation in 1965.

traveled freely through space, appearing today with a dominant wavelength corresponding to radio waves and a temperature of about 3 degrees above absolute zero. In 1965 Dicke's collaborators Roll and Wilkinson had just constructed an apparatus to search for their predicted cosmic radio waves when the radiation was discovered accidentally by Arno Penzias and Robert Wilson at Bell Laboratory in New Jersey. Penzias and Wilson were awarded the Nobel Prize for their discovery in 1978. To date, the most precise measurements of the cosmic background radiation have come from the Cosmic Background Explorer, a satellite launched in late 1989. This satellite has confirmed that the spectrum of the cosmic background radiation is extraordinarily close to that predicted by the big bang model.

The discovery in 1965 of the cosmic background radiation gave strong support to the idea that the universe was much hotter in the

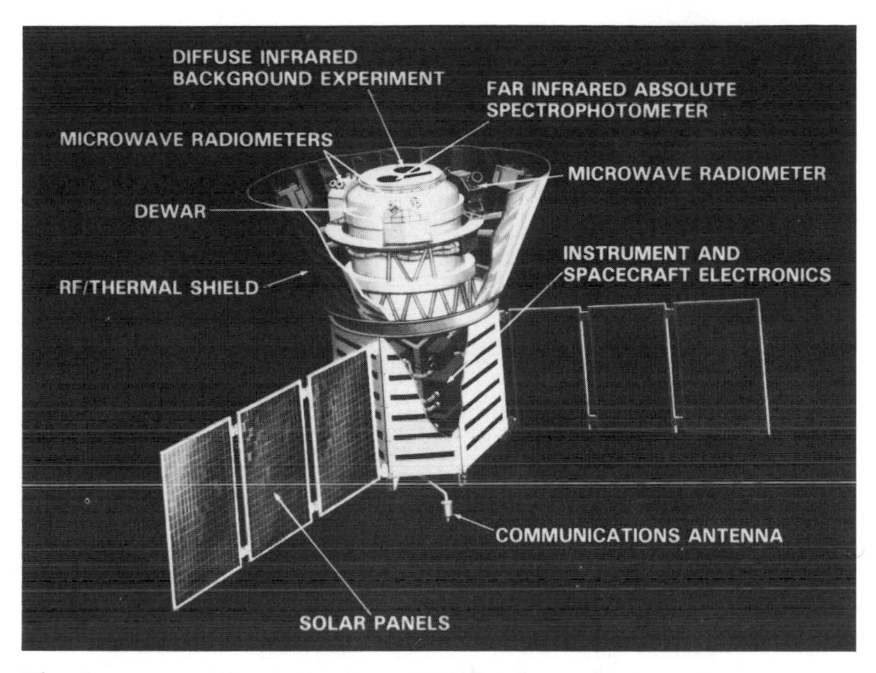

The Cosmic Background Explorer (COBE), a satellite launched in 1989 to study the cosmic background radiation.

past. Just as importantly, the observed cosmic radiation seems to confirm the hypothesis of large-scale homogeneity of the universe. The radiation has the same intensity from all directions in space—that is, it is isotropic. If we assume that we do not sit in an unusual place in the universe, then we can infer that the cosmic background radiation is isotropic at *any* location in the universe. This means that the universe was very homogeneous when the radiation last collided with matter, about 300,000 years after the big bang. If the universe were lumpy or of uneven temperature at such an epoch, the cosmic radiation would have scattered off the lumps in uneven intensities and directions and would not appear so uniform today. Because it has been traveling since the universe was only 300,000 years old, the cosmic radiation now detected has traveled much further than the distances to the visible galaxies, so it tells us about the smoothness of the universe on a much larger scale.

Observational confirmation of the large-scale homogeneity of the universe is vital for the standard big bang model and perhaps for all tractable cosmological models. As early as 1933, the British cosmologist Edward Arthur Milne suggested that the assumption of large-scale homogeneity might be logically necessary for any cosmological model. Milne named this assumption the cosmological principle, which has since become a starting point for most theoretical work in cosmology and has so far proven a necessary simplification for solving the difficult equations of the subject. If future observations cast doubt on the assumption of large-scale homogeneity, the gross features of the big bang model might still be correct, but the details would certainly be wrong.

Why didn't scientists immediately follow up on the original predictions of Alpher, Gamow, and Herman? (Indeed, Dicke was unaware of the earlier predictions and came to his conclusions completely independently.) There could be several reasons. The predicted cosmic radio waves were thought to be undetectable by the instruments of the 1950s. In addition, as Alpher and Herman recall, "Some scientists had a philosophical predilection toward a steady-state universe." In such a universe the temperature would always be what it is today and never hot enough to produce black-body radiation. Finally, most scientists in the 1940s and 1950s considered cosmology too speculative a subject to be taken seriously. There was practically no contact between theory and experiment.

C H A P T E R • F I V E

Other Cosmological Models

ONE variation of the big bang model, first discussed extensively by Richard Tolman of the California Institute of Technology in the early 1930s, is called the oscillating universe model. An oscillating universe is closed, but instead of ending after its collapse, it begins a new expansion, repeating the process of expansion and contraction through many cycles. If our universe were an oscillating universe, then it could be very much older than the estimated age of 10 billion years, which just measures the time since the last cycle of expansion began.

An apparent difficulty with this model results from the second law of thermodynamics, a basic law of physics that requires any isolated system to become more and more disordered until a state of maximum disorder is achieved. After many cycles, an oscillating universe would be expected to be much more chaotic than the universe we observe. Tolman was aware of this problem with oscillating models but argued that a state of maximum disorder might be impossible to define for the universe as a whole, thus rendering the objection uncertain. He concluded that "it would seem wise if we no longer assert that the principles of thermodynamics necessarily require a universe which was created at a finite time in the past and which is fated for stagnation and death in the future." Today, physicists are still not sure whether an oscillating universe would be theoretically ruled out by the second law of thermodynamics or whether the second law applies to the universe as a whole.

The oscillating universe model was in vogue in the late 1950s and early 1960s. In fact, it was a preference for an oscillating universe that led Robert Dicke to predict the existence of the cosmic background radiation. Dicke and his collaborators began their classic paper in the *Astrophysical Journal* in 1965 by pointing out that an oscillating universe, having existed for all time, "relieves us of the necessity of understanding the origin of matter at any finite time in the past." Taking such a model as a working hypothesis, Dicke then argued that if our universe had indeed gone through many cycles of expansion and contraction, its temperature would have to reach at least 10 billion degrees at each point of maximum contraction in order to disintegrate all the heavy elements created in stars during the previous cycle and to restore the matter of the universe to pure hydrogen. Otherwise, the nuclear reactions in stars would by now have converted most of the matter of the universe into heavy elements, contradicting the observations. Dicke then pointed out that at a temperature of 10 billion degrees, the reactions of subatomic particles would be sufficiently rapid to produce blackbody radiation. (In actuality, the production of such radiation does not *require* that the universe oscillate, only that the cosmic temperature have once been high enough.)

Beyond its possible violation of the second law of thermodynamics, the oscillating universe model lost favor in the 1960s when theoretical work by Roger Penrose and Stephen Hawking, both then at Cambridge University, showed that there was no plausible mechanism to produce oscillations. More specifically, Penrose and Hawking proved that the universe had to have originated at a density far higher than that previously contemplated and proposed for the "bounce" of each cycle of an oscillating universe. In actual fact, Penrose and Hawking's work did not rule out oscillating universes; it just eliminated all current models for oscillating universes and thus removed the scientific justification for belief in them.

In 1948 a group of restless young theoretical astrophysicists at

Cambridge University, not completely happy with the big bang model in any form and casting about for other possibilities, came up with the steady state model. This cosmological model, conceived of by Hermann Bondi, Thomas Gold, and Fred Hoyle, was not a variation of the big bang model. It proposed that the universe, on average, does not change in time. For example, the average density of matter does not change in time, and the temperature does not change in time. Philosophically, the steady state model of Bondi, Gold, and Hoyle was a rediscovery of Aristotle's static universe, adding a rigorous mathematical formulation and a knowledge of twentieth-century physics. The steady state model reconciles itself with Hubble's observations of the outward motion of galaxies by postulating that new matter and galaxies are continuously created throughout space, compensating for the spreading apart of individual galaxies and allowing the average number of galaxies per unit volume of space to remain constant. In this way, the universe maintains a steady state.

In their papers of 1948, Bondi, Gold, and Hoyle mention several reasons for proposing the steady state model. For one, they express dissatisfaction that the big bang model forces physicists to apply the laws of physics as observed today to a distant time in the past, when the conditions of the universe would have been far different. In the big bang model there is no way of knowing for sure whether the laws then were the same as now, yet no calculations can be made without assuming so. On the other hand, these scientists argue, a steady state universe is "compelling, for it is only in such a universe that there is any basis for the assumption that the laws of physics are constant." In the steady state model, the universe in the past was the same as the universe today. A second motivation for the steady state model was more quantitative: The rate of expansion of the universe as measured by the relatively uncertain techniques available in the 1940s translated to an estimated age for the universe of only 2 billion years; this was less than the geologically determined age for the earth. Some people considered this a problem for the big bang model. The steady state model also appealed to many scientists because, like the oscillating universe

Maarten Schmidt was born in Groningen, the Netherlands, in 1929, educated at the University of Groningen and the University of Leiden, and is now professor of astronomy at the California Institute of Technology. Schmidt is best known for his discovery of quasars. Schmidt likes to concentrate on a small number of topics and understand them well.

model, it eliminated the necessity of confronting the birth of the universe and all the uncertainties and incalculables attendant with that beginning. In the steady state model, the universe has no beginning and no end. Initial conditions do not have to be specified or accepted. Furthermore, some physicists and astronomers believed that the range of possibilities in such a universe would be much more limited than in the big bang model and thus easier to calculate. This was another attraction of the steady state model. Most physicists prefer theories that they can fully calculate. The steady state model was popular in the 1950s and early 1960s and considered the leading competitor to the big bang model and its variant oscillating universe model.

Today, almost all cosmologists believe that the steady state model has been ruled out. Besides the lack of evidence for the continuous creation of mass out of nothing or an explanation of how such a process could occur, the steady state model has been refuted by the discovery of the cosmic background radiation and other observations that suggest the universe was very different in the past. For example, the locations of certain astronomical objects called quasars (for "quasi-stellar radio sources") strongly suggest that the universe has changed over time. In 1963 Maarten Schmidt of the California Institute of Technology discovered these extremely luminous and distant pointlike sources of energy. A typical quasar lies 2 to 10 billion light years away and has the luminosity

Martin Rees was born in England in 1942, educated at Cambridge University, and is now professor of astronomy at Cambridge. Rees has worked on galaxy formation, galaxy clustering, and the origin of the cosmic background radiation, among other topics. He is known for his quickness, cleverness, and ability to keep several mutually inconsistent theories in his head at the same time.

of a hundred galaxies. In 1965 Martin Rees and Dennis Sciama of Cambridge University analyzed the data for the quasars known at that time and found that the number of quasars per volume of space increased with distance from the Milky Way. Since looking out to larger distances is equivalent to looking back in time, that meant that there were more quasars in the past. Rees and Sciama and others interpreted their result to contradict the steady state theory, which demands that the universe cannot change from one epoch to the next and hence cannot alter its population of quasars or galaxies or anything else.

Needless to say, both the oscillating universe model and the steady state model involve world views radically different from that of the "one-shot" big bang model. In the two former models, the universe has no beginning. As we will see in Chapter 10, the same idea has been incorporated in some very recent cosmological models.

CHAPTER • SIX

Difficulties with the Big Bang Model

DESPITE its successes, the big bang model has suffered from a number of problems. One troubling concern, ironically, is why the universe appears so uniform on the large scale. In particular, the incoming cosmic background radiation is remarkably uniform in all directions, varying in intensity by less than 1 part in 10,000 from different regions of the sky. The observed uniformity of this radiation indicates that the material gas of the universe had a nearly uniform density and temperature when the radiation last collided with it, about 300,000 years after the big bang. Although such uniformity and homogeneity is *assumed* in the big bang model, it still must be explained, or at least made plausible.

There are two possible explanations. Either the universe *began* with a high degree of homogeneity, or else any initial inhomogeneities eventually smoothed themselves out, much as hot and cold water in a bath tub will come to the same temperature by exchanging heat. However, heat exchange takes time. The regions of space that produced the cosmic radiation, when the universe was 300,000 years old, were about 50 million light years apart at that time—much too far apart to have had time since the big bang to exchange heat and homogenize. Thus, the second explanation simply doesn't work in the big bang model. The first explanation is considered unsatisfactory by some scientists because it seems to sweep the problem under the rug, relegating it to whatever unknown and currently uncalculable processes determined the initial conditions of the universe. Furthermore, it seems unlikely to many scientists that the universe would have been created so homogeneously. If nothing else, fluctuations in matter and energy arising from quantum processes, which will be discussed later, would naturally have produced lumpiness and irregularity in the very early universe.

The problem of accounting for the large-scale uniformity of the universe has been called the horizon problem. About each point in space one can picture a spherical volume that could have homogenized with the central point since the big bang. The outer edge of that sphere is called the horizon of the central point. Each point has its own sphere of homogenization and its own horizon. Since heat exchange, or any other homogenizing process, cannot travel faster than light, a point's horizon at any moment can extend no farther than the distance light can have traveled since the big bang. For example, the size of the horizon 300,000 years after the big bang was about 300,000 light years. Thus, 300,000 years after the big bang, each point in space could have homogenized with a spherical region around it extending only 300,000 light years. Beyond 300,000 light years from any given point, there would not have been time for light or heat or any other signal to have traveled across that distance since the beginning of the universe. The volume of space encompassed by the horizon is sometimes called the observable universe, since it is the region of space that can be seen by the central point at any given time. The observable universe today is a sphere 10 billion light years in radius. As time goes on, the horizon of any point grows, as does the size of the observable universe. The "horizon problem" arose because the smoothness of the cosmic background radiation suggests that different regions of the universe separated by *more* than each other's horizon (about 150 times more) nevertheless appear to have exchanged heat.

The horizon problem seems to have been first clearly stated in print in 1969 by Charles Misner of the University of Maryland. Although Einstein simply assumed large-scale homogeneity, he would have had no trouble explaining how it came about. Since

Charles Misner was born in Jackson, Michigan, in 1932, educated at the University of Notre Dame and Princeton University, and is now professor of physics at the University of Maryland, College Park. Misner was the first to pose the horizon problem and pioneered a series of studies aimed at explaining the properties of the universe in terms of known physical processes, rather than ad hoc initial conditions and assumptions.

Einstein also assumed that the universe had existed forever, there would have been plenty of time for any two regions, arbitrarily far apart, to have exchanged heat and homogenized. Such an explanation does not work in the big bang model.

A closely related, and more controversial, issue is the so-called flatness problem: Why should the universe today be so near the boundary between open and closed, that is, so nearly flat? In other words, why is the measured value of omega—the ratio of the cosmic mass density to the critical density needed to close the universe—today so close to 1? It follows from the big bang model that as time goes on, omega should differ more and more from 1, unless it started out exactly 1, in which case it remains 1. In an open universe, omega begins smaller than 1 and gets smaller and smaller in time; in a closed universe, omega begins larger than 1 and gets larger and larger.

Omega is analogous to the ratio of gravitational energy to kinetic energy of motion of a rock thrown upward from the earth. If the rock is launched with precisely the critical speed, that ratio will start out 1 and remain 1. If the rock is thrown with less than the critical speed, the ratio will start out greater than 1 and continuously increase, becoming infinite just when the rock reaches maximum height and is about to fall back to earth. At this point, the rock has zero speed, its kinetic energy of motion is zero, and the ratio of gravitational energy to kinetic energy is therefore infinite. In contrast, if the rock is thrown with more than the critical speed, the

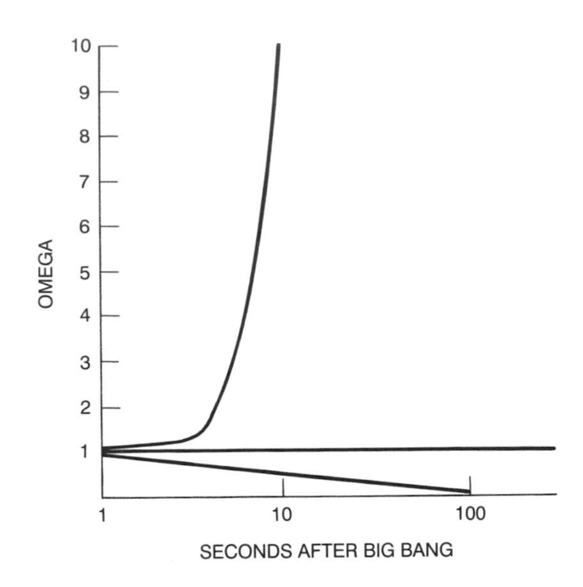

Behavior of omega versus time for three representative initial values: bigger than 1, 1, and less than 1.

ratio will start out less than 1 and continuously decrease, approaching zero as the rock escapes the earth's gravity altogether and journeys off into outer space. Finding the cosmic omega so close to 1 today, so long after the big bang, is analogous to sighting the rock a long time after it was launched—very, very far from earth—and finding its gravitational energy and kinetic energy of motion almost equal. Such an event is highly unlikely because it would require the two energies at launch to have been balanced to extraordinary precision. For example, if a rock is thrown upward with an initial ratio of energies of 0.75, that ratio will have dropped to 0.1 by the time the rock has reached a distance of 27 earth radii; for an initial ratio of 0.9, the ratio will have dropped to 0.1 at a distance of 81 earth radii. For the rock to reach a million (10^6) times its initial distance before the ratio falls to 0.1, the initial ratio had to be 0.999991. The numbers behave similarly for ratios larger than 1.

Physicists believe that the initial conditions of the universe were set when the universe was about 10^{-43} seconds old. In order for the value of omega to still remain between 0.1 and 10.0 today, after 10 billion years and after the universe has expanded to 10^{30} times its initial size at "launch," the initial value of omega had to lie between $1 - 10^{-59}$ and $1 + 10^{-59}$. Equivalently, the kinetic energy and the gravitational energy of the universe had to be initially equal to one part in 10^{59}. What physical processes could have set so fine a balance? And there is another puzzle. If the gravitational and kinetic energies are not exactly equal today, why are they becoming unbalanced at this particular moment in cosmic time, just when *Homo sapiens* happened to arrive?

The flatness problem seems to have been first clearly posed and put into print by Robert Dicke in 1969. Although several British cosmologists, including Brandon Carter and Stephen Hawking, independently noted the problem shortly thereafter, it was not widely known or understood until it was restated in an article by Dicke and Peebles in 1979.

There is a broad range of attitudes about the flatness problem. Some scientists consider the initial value of omega to be an accidental property of our universe, a value that should be accepted as a given, and they see no validity in the flatness problem. For this group of cosmologists, the flatness problem is a nonproblem, an issue lying beyond the domain of science. Others agree with Dicke and Peebles that the required initial conditions seem too special to be accidental and that some deeper physical explanation is required. Among this latter group are scientists who say that for some reason the initial gravitational and kinetic energies must have been balanced *exactly*. Omega was and is exactly 1. This view requires the existence of a huge quantity of undetected mass. Since we have observed only enough mass per volume of space to make omega equal to about 0.1, the belief that omega is really exactly 1 requires that there be, on average, about 10 times as much mass as has been observed in every cubic light year of space.

Before 1980 most cosmologists put aside or paid little attention to the flatness problem. After an influential modification of the big bang model called the inflationary universe model, which gave a natural solution to the flatness problem and which will be discussed in Chapter 10, many scientists considered the flatness problem to be important. The controversial status of the flatness problem is evident in Alan Guth's seminal paper on the inflationary universe model, where an appendix is devoted to convincing skeptics that the problem should not be dismissed. Even today there is still no consensus on the meaning or depth of the flatness problem.

Another old cosmological problem has been the lack of a good explanation for the average number of radiation particles, called photons, relative to the number of baryons. (Examples of baryons are protons and neutrons, which make up the nuclei of atoms.) We do not know the total number of photons or baryons in the universe, but the ratio of these numbers can be estimated by counting photons and baryons in a large volume of space and then assuming that volume is typical of the universe as a whole. The measured ratio is about a billion photons for every baryon. What makes this number fundamental is that it should be constant in time, according to the theory. It is a fixed property of the universe. But what determined its value? As in the flatness problem, some scientists have invoked the accident of initial conditions to explain why the photon-to-baryon ratio has the value it does, saying, in effect, that the number is a billion now because it was a billion then. Other scientists believe that this number should be calculable from basic principles. The big bang model itself does not require the photon-to-baryon ratio to have any particular value, just as it does not require the initial value of omega to have been anything in particular.

Finally, a problem whose importance has been appreciated only recently concerns the entropy of the universe. In the nineteenth century, scientists discovered the second law of thermodynamics, which states that any isolated physical system subjected to random disturbances will naturally become more disordered in time, that is, will naturally increase its entropy. Entropy is a quantitative measure of the disorder of a physical system. For example, a deck of cards arranged with all the cards of each suit together is very organized. Such a well-ordered deck is said to have low entropy. A deck that has been shuffled many times, with its cards in random positions, is said to have high entropy. Intuitively, the second law

Roger Penrose was born in Colchester, England, in 1931, educated at University College London and Cambridge University, and is now professor of mathematics at Oxford University. Among Penrose's contributions to cosmology is work with Hawking on the cosmological singularity theorems and his statement of the entropy problem in cosmology. Penrose is one of the most mathematically sophisticated people ever to have worked in general relativity and cosmology and is renowned for his discovery of Penrose tiles, which are two geometric shapes that can be fit together over and over to completely cover a 2-dimensional plane, without repeating any pattern.

of thermodynamics makes sense. If you start with a deck of cards arranged by suit and number and drop it on the floor, the odds are great that the regathered cards will not be arranged in good order. On the other hand, if you start with a randomly ordered deck and shuffle it 10 times, the chances are extremely small that the resulting cards will be arranged in ascending order. Similarly, eggs often break but never reform; skywriting fades but never comes back; unattended rooms gather dust but do not get clean. Any isolated system evolves in a one-way direction from order to disorder.

In a series of papers beginning in 1974, Roger Penrose of Oxford University applied the second law of thermodynamics to the universe as a whole. More specifically, Penrose estimated the entropy or disorderliness of the observable universe and found it to be fantastically small compared with what it theoretically might be—for example, if much of the cosmic mass were in the form of a huge black hole, rather than in galaxies. If one traces cosmic evolution backward in time, the second law of thermodynamics decrees that the universe began with an even greater degree of order—even lower entropy. Penrose and others find it mysterious that the universe was created in such a highly ordered condition—as if we were dealt a royal flush—and believe that any successful theory of cosmology should ultimately explain this entropy problem. The big bang model, in its present form, does not. Indeed, the big bang model says nothing about the initial conditions of the universe.

C H A P T E R • S E V E N

Large-Scale Structure and Dark Matter

THE basic assumption of homogeneity that underlies the big bang model is obviously not true nearby. The universe nearby is not evenly filled with a smooth and featureless fluid. Rather, it is lumpy. Matter clumps into galaxies and galaxies clump into clusters of galaxies and so on. Astronomers called such clustering of galaxies "structures." Structures of various sizes abound, and astronomers want to understand the nature of these structures and how they were formed. Until such understanding, it will be hard to decide whether the observed inhomogeneities are simply details in the standard view or hints of a radically different picture.

One of the first twentieth-century scientists to suggest a nonuniform distribution of mass in the universe was C. V. L. Charlier, who considered the possibility of a hierarchical cosmos, in which the structure and average density would change as one went to larger and larger scales. In 1933 Harlow Shapley of the Harvard College Observatory commented that the observed irregularities in the positions of galaxies were too pronounced to be accidental groupings in a basically smooth background and suggested some "evolutionary tendency in the metagalactic system." Fritz Zwicky of the California Institute of Technology suggested in 1938 that clusters of galaxies, of about 10 million light years in size, be considered the basic units of matter in the universe.

In 1953 Gérard de Vaucouleurs, then at the Australian National Observatory, discovered that the galaxies within about 200 million light years of the Virgo cluster of galaxies, which includes the Milky Way, were mostly confined to a giant disk. He called this large congregation of galaxies a supercluster of galaxies. With this discovery and subsequent work, de Vaucouleurs challenged the assumption of homogeneiety in the universe and, in fact, proposed a hierarchical universe, in which small structures are part of larger structures, which are part of even larger structures, continuing up indefinitely. In this picture, not only is the universe inhomogeneous, but an average density of matter cannot be defined. The larger the volume one takes to measure the density, the smaller the density. If such a model were correct, few conclusions about the overall behavior of the universe could be drawn from measurements in our vicinity.

Further evidence for inhomogeneities has come in recent years. In studies of large groups of galaxies in 1975, G. Chincarini and H. J. Rood discovered lumpiness in the distribution of matter over distances of roughly 20 million light years. At a symposium of the International Astronomical Union in 1977, W. G. Tifft and S. A. Gregory and, independently, M. Joeveer and J. Einasto of Estonia reported their observation of clusters and chains of galaxies and "voids" with no galaxies at all, extending over distances of several hundred million light years. In 1978 Gregory and L. A. Thompson found evidence for a large congregation of galaxies, called the Coma supercluster, with relatively empty space around it. In 1981 Robert Kirshner, August Oemler, Jr., Paul Schechter, and Stephen Shectman found a huge void in space, in the direction of the constellation Boötes, with a diameter of about 100 million light years. It appears that few galaxies inhabit this vast empty space. (For comparison, in our own cosmic neighborhood one can find a galaxy every few million light years or so.) About the same time, Gregory, Thompson, and Tifft documented a large number of galaxies apparently arranged in a long chain extending about 100 million light years, called the Perseus–Pisces chain, previously identified by Joeveer and Einasto. The Perseus–Pisces chain of

Gérard de Vaucouleurs was born in Paris in 1918, educated at the University of Paris, and is now professor of astronomy at the University of Texas at Austin. De Vaucouleurs' contributions to cosmology include the identification of the local supercluster, compilation of reference catalogues of galaxies, advocacy of a hierarchical distribution of cosmic matter, and new measurements of the rate of expansion of the universe. De Vaucouleurs is known for taking controversial positions that turn out to be true.

galaxies was more carefully defined and studied in 1985 by H. P. Haynes and R. Giovanelli of Cornell University.

In 1986 observations by Margaret Geller, John Huchra, and Valerie de Lapparent of the Harvard-Smithsonian Center for Astrophysics revealed that the galaxies in a certain region of space appear to be located on the surfaces of bubble-like structures of about 100 million light years in size, with voids inside the bubbles. These new observations were among the first to show the 3-dimensional locations of a large, connected sample of galaxies (1,100 galaxies in the 1986 survey). Extending their survey to several thousand galaxies, Geller and Huchra reported in 1989 evidence for a "wall" of galaxies stretching at least 500 million light years in length. These recent 3-dimensional maps of large samples of galaxies, called redshift surveys, have been made possible by advances in technology that allow the redshifts of galaxies to be measured in fast and automated procedures. As mentioned previously, under the assumption that the universe is approximately homogeneous and uniformly expanding, the redshift of a galaxy translates into an approximate distance, thus providing the elusive third dimension for the position of a galaxy. In the coming decade, astronomers hope to initiate redshift surveys of 1 million galaxies, using telescopes specifically dedicated to that job.

Very recent work by a collaboration of Queen Mary and Westfield College in England, the University of Durham in England,

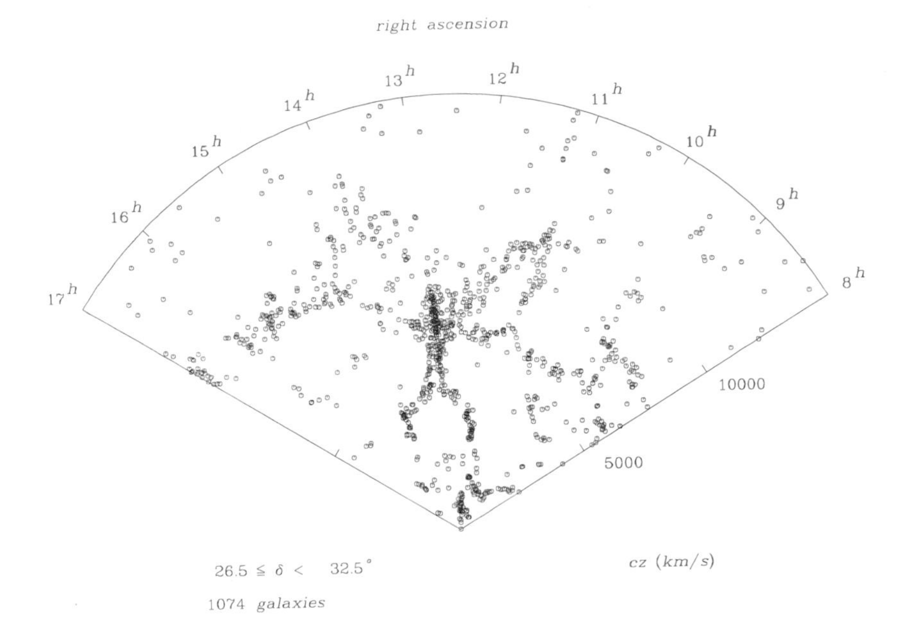

An early Center for Astrophysics redshift survey (1986), showing the positions of 1,074 galaxies in a thin pie-shaped slice through space. Each circle is a galaxy. Our position is at the bottom of the wedge. The radial direction directly measures recessional speed, which corresponds to radial distance in a homogeneous universe. The farthest galaxy surveyed here is approximately 500 million light years away.

the University of Oxford, and the University of Toronto—to be discussed later—suggests that the cosmic structures found in a few selected regions of space may be typical. It also seems clear that structures of *some* kind have usually been found at the largest possible scale in each survey of galaxies; that is, a survey that looks over a region of 100 million light years usually finds some chain or disk or *absence* of galaxies extending roughly a 100 million light years in size; a survey of a 200-million-light-year region finds structures of 200 million light years, and so on. Galaxy surveys are now in progress that extend out to a billion light years. It remains to be seen whether these surveys will also show recognizable patterns and structures over such large regions.

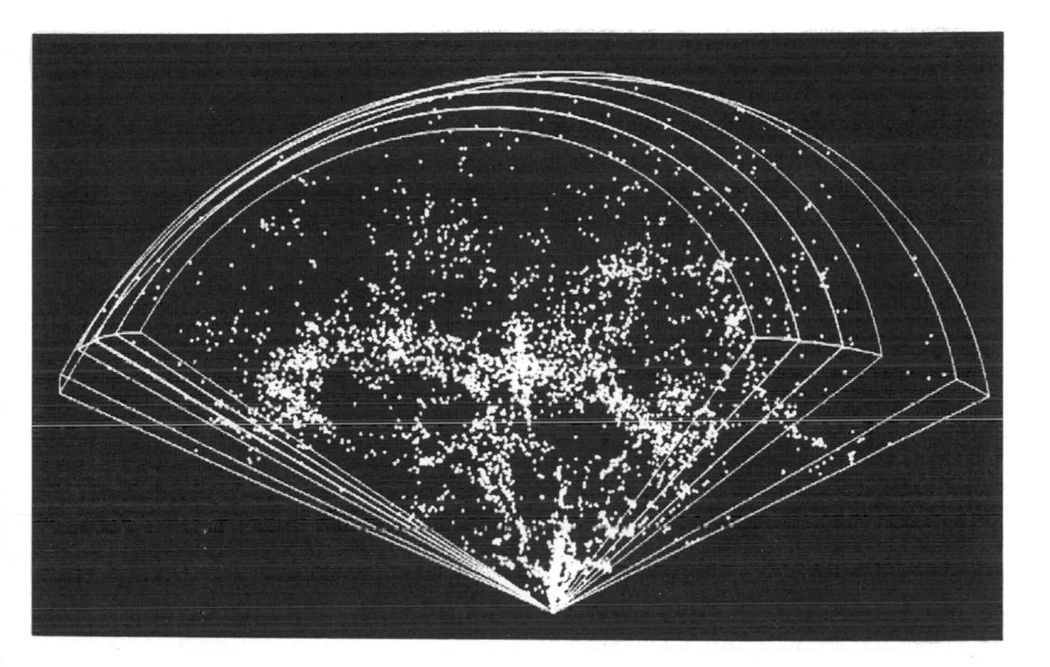

A larger Center for Astrophysics survey (1989), compiled by Margaret Geller and John Huchra, showing the positions of 3,962 galaxies in several touching wedges in space. The farthest galaxy is about 500 million light years away. The nearly continuous horizontal bunch of galaxies stretching across the diagram has been called the "Great Wall" and is the largest coherent structure of galaxies yet observed.

Some cosmologists are worried that the observed inhomogeneities in regions of several hundred million light years might extend *indefinitely,* to ever larger sizes, and threaten the foundations of the big bang model—particularly its assumption of large-scale homogeneity. Challenging this view of a hierarchy of structures on all scales are recent redshift surveys by T. J. Broadhurst and R. S. Ellis of the University of Durham in England, David C. Koo of the Lick Observatory, Richard Kron of the University of Chicago, and Alex S. Szalay of Johns Hopkins University and Eotvos Lorand University of Budapest. These surveys are essentially 1-dimensional; they compute the redshifts of galaxies only along one outward line through space and find the positions of galaxies along

Margaret Geller was born in Ithaca, New York, in 1947, educated at the University of California and Princeton University, and is now professor of astronomy at Harvard and an astrophysicist at the Smithsonian Astrophysical Observatory. With John Huchra, Geller has led the Center for Astrophysics redshift survey and found evidence that some galaxies appear to be located on the surfaces of bubble-like structures, about 100 million light years in diameter. Geller attributes her strong capacity to visualize in 3 dimensions to childhood training by her father. Margaret Geller was a student of Jim Peebles.

that line, like beads on a string. To compensate for their lack of breadth, these "pencil beam" surveys go extremely deep, out to several billion light years. The pencil beam surveys suggest a highly regular and organized, lattice-like cosmic structure, with galaxies bunched together every 400 million light years. While such regularity is probably not general, according to this picture the universe would appear homogeneous when averaged over regions much larger than 400 million light years. Larger and broader studies are badly needed to test these intriguing and important new results.

Many scientists, led by James Peebles of Princeton University, are persuaded by surveys of very faint radio-emitting objects and the nearly uniform cosmic X-ray emission that the matter of the universe becomes smooth when averaged over several billion light years or more. That could be the distance over which the individual grains of sand on the beach would no longer be seen. Most cosmologists have confidence that the universe must be homogeneous when viewed on scales of 10 billion light years, since the cosmic background radiation is smooth and comes from such distances. If in the future we find filaments and bubbles and voids with sizes of a few billion light years, several times larger than those now mapped, then there would be a direct contradiction with the uniformity of matter implied by the cosmic background radiation. The big bang model could be thrown into crisis. At present, many

John Huchra was born in Jersey City, New Jersey, in 1948, educated at the Massachusetts Institute of Technology and the California Institute of Technology, and is now professor of astronomy at Harvard and an astrophysicist at the Smithsonian Astrophysical Observatory. With Margaret Geller, Huchra has led the Center for Astrophysics redshift survey and found evidence that some galaxies appear to be located on the surfaces of bubble-like structures, about 100 million light years in diameter. Huchra is an observer's observer and has more hands-on experience with telescopes than almost anyone his age.

cosmologists feel that the observed inhomogeneities in the distribution of galaxies certainly have implications for the formation of galaxies and clusters of galaxies but do not yet conflict with the big bang model and its assumption of homogeneity on very large scales. In any case, the large-scale structures must be reckoned with.

Related to observations of cosmic structure are the peculiar velocities of galaxies, that is, velocities that depart from the recessional velocity expected in a perfectly smooth and uniformly expanding universe. As mentioned earlier, when the material of the universe is lumpy, the outward speed of a galaxy is no longer strictly proportional to its distance. The motions of galaxies are altered by the irregularity of the gravity they feel. Thus, peculiar velocities of galaxies are an indirect sign of inhomogeneities in the distribution of cosmic mass. To measure the peculiar velocity of a galaxy, a scientist must know its distance as well as its redshift. The redshift to the galaxy gives only its total velocity. As previously discussed, to know how much of this is "normal" and how much "peculiar," the normal velocity at that distance—that is, the distance multiplied by the rate of expansion of the universe (Hubble constant)—must be known. Assuming the Hubble constant is already known, the distance to the galaxy must be known. Thus, measurements of distance are crucial in all studies of peculiar veloc-

Sandra Faber was born in Boston in 1944, educated at Swarthmore College and Harvard University, and is now professor of astronomy at the University of California at Santa Cruz. Among her contributions to cosmology, Faber discovered a new method for determining the distances to galaxies, and she was one of the Seven Samurai who discovered the large-scale motion of galaxies toward the Great Attractor. Sandra Faber was a student of Vera Rubin's.

ities. It cannot be assumed that distance is proportional to redshift—such an assumption is equivalent to the assumption of homogeneity, which is precisely what is being tested.

Work on the peculiar velocities of galaxies was pioneered by Vera Rubin of the Carnegie Institution of Washington, as early as 1951, and continued by her and N. Thonnard, W. K. Ford, Jr., and M. S. Roberts in the mid 1970s. In 1987, using improved methods for measuring cosmic distances, David Burstein of Arizona State University, Roger Davies of the National Optical Astronomy Observatories, Alan Dressler of the Carnegie Institution of Washington, Sandra Faber of the University of California at Santa Cruz, Donald Lynden-Bell of Cambridge University, Robert J. Terlevich of the Royal Greenwich Observatory, and Gary Wegner of Dartmouth College (dubbed the Seven Samurai) found evidence that a large group of galaxies within about 200 million light years of us have been substantially deflected in their motions, as if attracted by some large mass. The speed of this peculiar motion is about 10 percent of the expansion speed at that distance, and the big lump of mass believed responsible has been called the Great Attractor. The Great Attractor, which is clearly a gross inhomogeneity in the cosmic mass distribution, appears to be a concentration of mass extending over several hundred million light years.

Significant peculiar motions, such as found by the Seven Samurai, could complicate an accurate measurement of the overall rate

Vera Rubin was born in 1928 and raised in the Washington, D.C., area, educated at Vassar, Cornell, and Georgetown University, and is now a staff member of the Department of Terrestrial Magnetism of the Carnegie Institution of Washington. Rubin was one of the first to document the peculiar motions of galaxies. She also did one of the first observational studies clearly indicating the presence of dark matter. In 1965, Rubin became the first woman permitted to observe at the Palomar Observatory.

Alan Dressler was born in Cincinnati, Ohio, in 1948, educated at the University of California at Berkeley and at the University of California at Santa Cruz, and is now on the staff of the Mt. Wilson and Las Campanas Observatories of the Carnegie Institution of Washington. His research includes work on the formation and evolution of galaxies, the large-scale structure of the universe, and motions of galaxies. Dressler is one of the Seven Samurai.

of expansion of the universe, as well as being difficult to understand in their own right. On the other hand, peculiar velocities provide a unique tool for mapping the mass inhomogeneities of the universe. Beginning in 1989, Edmund Bertschinger of the Massachusetts Institute of Technology, Avishai Dekel of the Hebrew University in Jerusalem, and their collaborators have developed a new theoretical method for inferring the distribution of cosmic mass in a region of space, given the observed peculiar velocities of galaxies in that region. The Bertschinger–Dekel method assumes that the observed peculiar velocities are caused by the irregular gravity of mass inhomogeneities. Of particular interest is the question of whether the inferred distribution of mass coincides with the mass that we see in the galaxies themselves. At least some of the peculiar velocities of galaxies can be explained by the observed inhomogeneities of matter—the galaxy clusters, chains, walls, and voids—plus gravity. Any large cluster of galaxies will gravitationally attract other galaxies in its vicinity, and the resulting motions show up as peculiar velocities. The question is whether the observed inhomogeneities of matter can *fully* account for the observed peculiar velocities. If not, then there must be some previously unobserved inhomogeneities, such as the Great Attractor, or there must be some form of matter that cannot be seen, or additional forces besides gravity must be working. Any of these possibilities would send theorists back to the drawing board.

In addition to new theoretical work, new observational studies of peculiar velocities are under way. As in the redshift surveys, these new studies need to cover larger regions of the sky and need to penetrate to farther distances. The goal for the near future is to map the peculiar velocities of about 15,000 galaxies out to a distance of about 300 million light years. As mentioned earlier, independent measurements of cosmic distances are critical for these studies, and for all studies of large-scale structure.

An obstacle to understanding the distribution of mass in the universe and motions of galaxies is that about 90 percent of the detected mass in the universe is invisible. It emits no radiation of any kind—not optical light, nor radio waves, nor infrared, nor ultraviolet, nor X-rays. It is truly invisible. This detected but unseen matter is called dark matter. We know that dark matter exists, because we have detected its gravitational effects on the stars and galaxies that we see, but we have little idea what it is. The problem was noticed in 1933 by the Swiss-American astronomer Fritz Zwicky. Zwicky was able to estimate the mass of a cluster of galaxies in orbit about one another by measuring the amount of gravity needed to hold the cluster together. He discovered that the total mass thus inferred was about 20 times what could be accounted for by the visible stars in the cluster.

Zwicky's startling discovery was not appreciated nor reckoned with for many years. In 1973 Jeremiah Ostriker and James Peebles of Princeton did theoretical analyses suggesting that the amount of visible matter in the disks of typical rotating galaxies was not

Jeremiah Ostriker was born in New York City in 1937, educated at Harvard University and the University of Chicago, and is now professor of astrophysical science at Princeton. Ostriker's contributions to cosmology include the prediction of the existence of dark matter and the proposal that huge explosions, and the subsequent compression of gas, may play a role in the formation of galaxies. Ostriker is witty, quick, and able to leap from one theory to the next.

sufficient to keep those galaxies from flying apart or drastically changing shape. The observation of such galaxies quietly spinning around therefore suggested that their outer regions were filled with a halo of unseen mass, comparable to the visible mass of the disk, holding the galaxy together by its gravity. The next year Ostriker, Peebles, and Amos Yahil compiled observations of inferred mass and visible mass of various astronomical systems, from individual galaxies to large clusters of galaxies, and claimed that 90–95 percent of the mass of the universe is in an invisible form. J. Einasto, A. Kaasik, and E. Saar independently arrived at the same conclusion. Four years later Vera Rubin and colleagues at the Carnegie Institution of Washington and Albert Bosma of the University of Gronigen estimated the mass of several galaxies by measuring the speed at which gas orbits around the center of each galaxy. These researchers found direct and compelling evidence for about 5 times more mass in spiral galaxies than can be accounted for in visible stars. Similar measurements have been repeated for groups of galaxies orbiting one another and have found 10 times as much dark matter as visible matter.

It is important to distinguish between this dark matter, which has been *detected* through gravitational studies but not seen, and the unseen mass *hypothesized* by scientists who believe that omega equals 1. The latter we will call "missing mass." To review, the

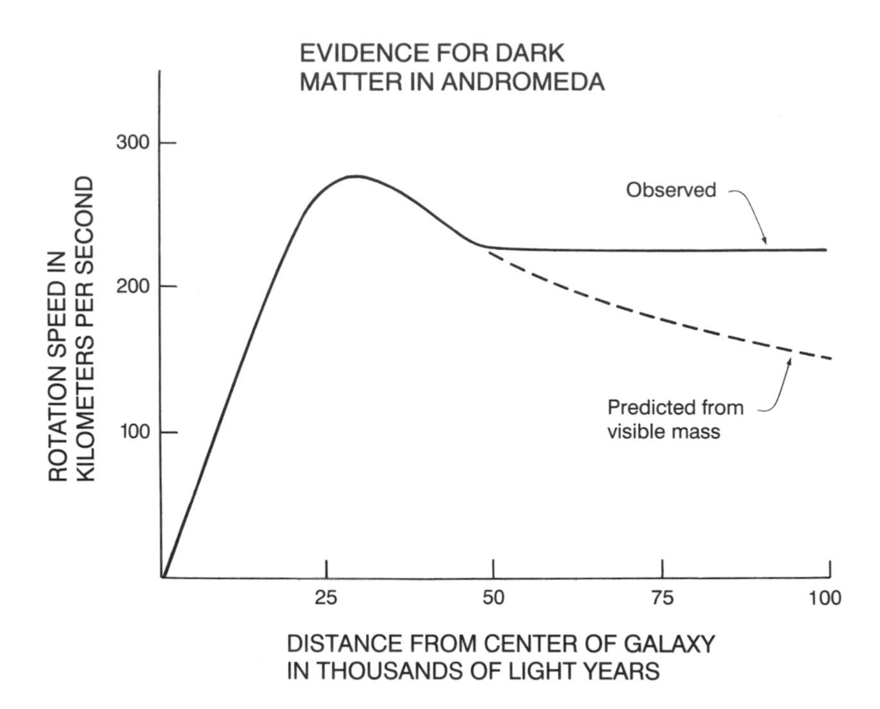

light-emitting, visible mass provides only enough density of matter to make omega equal to 0.01. Including the gravitationally detected but invisible mass—the dark matter—leads to an omega of about 0.1. A value of 1.0 for omega requires the existence of 10 times more mass, not only unseen but undetected altogether—the missing mass. We know that dark matter exists. The missing mass may or may not.

What is the nature of dark matter? Does it consist of numerous planets, or of collapsed stars called black holes, or of subatomic particles that interact with other matter only through gravity, or perhaps some new, as yet undiscovered type of subatomic particle? Depending on what it is, the dark matter could alter our theories of subatomic particles or of the formation of galaxies; it must be identified for any secure understanding of the cosmos. In recent years astronomers have been deeply disturbed to realize that the luminous matter they have been staring at and pondering for centuries makes up a mere tenth of the inventory.

It is not just the unknown identity of dark matter that causes concern. Its quantity and arrangement in space are also uncertain, foiling attempts to understand why the luminous mass is arranged as it is. A careful reconciliation of the peculiar velocities of galaxies with the *observed* inhomogeneities in luminous matter should reveal the presence of dark matter, which contributes to peculiar velocities through its gravitational effects. Preliminary results of Bertschinger, Dekel, Faber, and others, using the methods mentioned earlier, suggest that dark matter may be distributed in the same way as visible matter. Detailed maps of the positions and motions of galaxies, over large scales, will yield better maps of the location of dark matter.

There are other ways to probe dark matter. One of the most recent new techniques, and potentially very important, makes use of the "gravitational lens" phenomenon. When Einstein published his new theory of gravity, he pointed out that light, like matter, should be affected by gravity. Thus, as light from a distant astronomical object, such as a quasar, travels toward the earth, that light should be deflected by any mass lying between here and there. The intervening mass can act as a lens, distorting and splitting the image of the quasar. Even if the intervening mass is totally invisible, its gravitational effects are not. By carefully analyzing the distortions of quasar images, theoretical astronomers can reconstruct many of the properties of the intervening gravitational lens, including its distribution in space and total mass. Gravitational lenses were first discovered in 1979; about a dozen have been found since that time. Very recently, Anthony Tyson of AT&T Bell Laboratories and his collaborators have used the gravitational lens phenomenon to map the distribution of dark matter in clusters of galaxies. Similar methods will be a powerful tool in the coming decade.

Dark matter could be something mundane, like large planets, and this possibility should be ruled out before more exotic options are explored. Some astronomers have proposed that the dark matter consists of large planets with masses between a thousandth and a tenth the mass of our sun. Such objects should have enough heat generated by their slow contraction to emit a low intensity of infrared radiation, which is radiation with wavelengths longer than those of visible light. A highly sensitive infrared telescope might detect such massive planets and is being planned for the 1990s.

Putting aside for the moment the uncertainty of dark matter, a number of different theories have been proposed for the formation of galaxies and their distribution in space. In any such theory, two things must be specified: the initial positions and motions of lumps in the otherwise smooth distribution of cosmic mass, and the subsequent forces acting on the lumps. Scientists have traditionally assumed that it is mainly the force of gravity that acts on the initial lumps.

One of the earliest pictures, called the gravitational hierarchy model, was first sketched out by the British astronomer James Jeans in 1902, for a static universe, and then modified for an expanding universe by Georges Lemaître in 1933. Jeans and Lemaître supposed that the mass in the early universe was nearly uniform but was bunched up very slightly here and there, like small ripples on the surface of a pond. The origin of these ripples they left for a later theory. In a place where the mass was bunched up, there would be slightly stronger gravity. This would cause nearby mass to bunch up more, attracting more surrounding gas. Gravity would get stronger, and the process would continue until a strong concentration of mass had formed. Small initial ripples and lumps would eventually produce single galaxies, larger ones would produce groups and clusters of galaxies, and so on, with a hierarchy of structures. The locations of galaxies in space today would then be explained by the locations and sizes of the initial ripples, together with the subsequent action of gravity. In recent years, beginning with the work of James Peebles in the mid 1960s, this theory has been made quantitative, but it does require specification of the sizes and strengths of the initial ripples.

Modern theoretical investigations of the gravitational hierarchy model have used computers to simulate the growth of lumps in an expanding universe. In such a simulation, called an N-body simulation, 10,000 to 10 million mass points, each representing a galaxy or portion of a galaxy, are placed at initial positions, given an initial

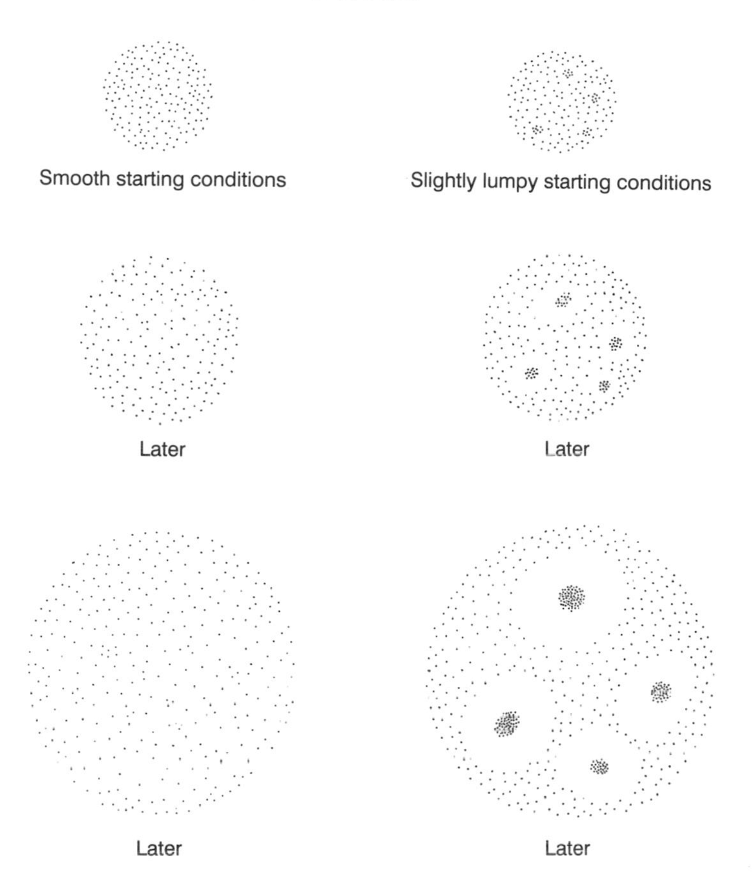

outward velocity corresponding to the expansion of the universe, and then allowed to interact via their mutual gravity. Dark matter and missing matter may also be added, making up some assumed fraction of the total mass and distributed in some assumed way. The hypothetical galaxies fly around the computer screen, gravitate toward each other, and form clumps and wisps and voids. Recently, R. Y. Cen and Jeremiah Ostriker of Princeton and collaborators have added the effects of gas pressure to computer simulations of how galaxies cluster together under their mutual gravity. These effects—which arise in part from the fact that individual galaxies are not points of mass but have finite extension in space—are important over distances of 100 million light years and smaller.

In 1967 Joseph Silk, then at Harvard, calculated that lumps of matter initially smaller than about a thousand times the visible mass of a galaxy might not be able to hold together under the effects of radiation. This result was incorporated into the "pancake" model, developed in the early 1970s by Y. B. Zel'dovich, A. G. Doroshkevich, and others in Moscow. In this model, the first lumps of mass to begin growing were very large. There would have been many such lumps, of course. As each lump cooled, it would collapse under its own weight, with the collapse tending to be fastest in one direction. The result would be a thin "pancake" of gas, which would then fragment into many small pieces. Each piece would be an individual galaxy. In this picture, galaxies would tend to be distributed in sheets, following the shape of their parent gas cloud.

The gravitational hierarchy proposal is a bottom-up model for the formation of cosmic structures, with small lumps of matter condensing first and then growing larger and larger. By contrast, the pancake model is a top-down model, with large aggregates of matter forming first and then splitting apart into smaller structures.

Any convincing theory of structure formation must explain the observed distribution of galaxies. In particular, cosmologists must explain why many galaxies are located in relatively thin sheets. (Here, thin means that the width is far less than the height or breadth, although that width might be a million light years across.) The big bang model assumes that gravity is the major force in determining the evolution and structure of the universe. And conventional wisdom has long held that gravity acting by itself produces smoothly varying features in the locations of masses, with comparable widths, heights, and breadths of any groupings of galaxies. According to this view, additional physics or special initial conditions are needed to produce sharp features in the mass distribution, such as linear strings and thin sheets of galaxies.

Conventional wisdom has been challenged by recent computer simulations, which show that sharp features can indeed occur if the

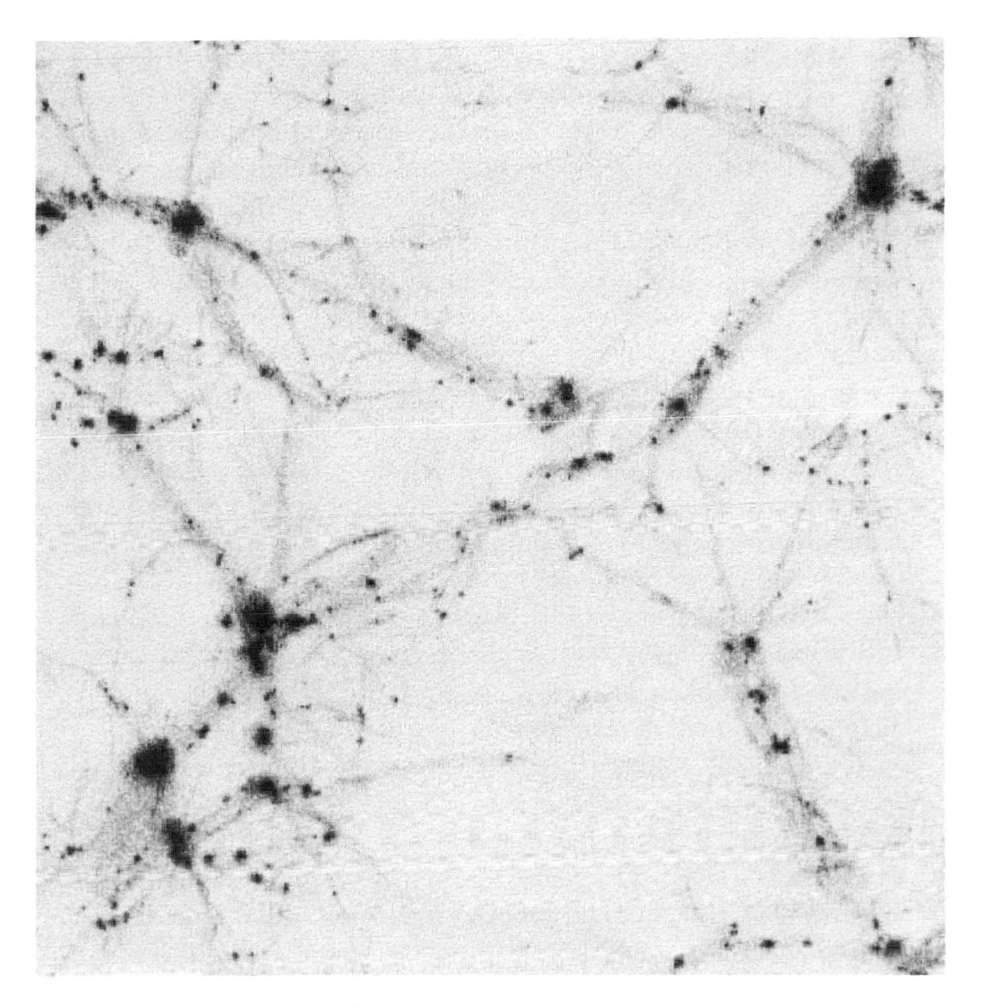

Computer simulations of cosmic mass clustering, done by James M. Gelb and Edmund Bertschinger of MIT. The calculations assume the cold dark matter model, which includes a value of 1 for omega. The figure shows a snapshot after significant evolution and clustering has occurred and is supposed to correspond to the current epoch. There are about 17 million particles, each about 0.001 galaxy mass. Each visible small dot is 1 galaxy; each big smudge is several hundred galaxies. The region shown covers a maximum distance of about 150 million light years. The calculations were done on an IBM 3090 supercomputer at the Cornell National Supercomputer Facility.

initial inhomogeneities are sufficiently pronounced in small patches and short distances. These new calculations, by Changbom Park and Richard Gott at Princeton University, Edmund Bertschinger and James Gelb at MIT, and Jens Villumsen at Ohio State University, use several million mass points in a version of the gravitational hierarchy model called the cold dark matter model. (The largest computer simulation to date, by Gelb and Bertschinger, uses 17 million particles and was performed on an IBM 3090 supercomputer.)

For the last decade the cold dark model has been the leading contender to explain the formation of galaxies and other large-scale structures. The model is based on the inflationary universe model (to be discussed in chapter 10), which requires that omega be 1 and which specifies the initial inhomogeneities in the infant universe. The model gets its name from the fact that the dark matter particles—whatever they are—are assumed to be slowly moving, that is cold, and thus easily deflected by gravity. Many theorists working on the problem of large-scale structure in the universe have adapted the cold dark matter model as a starting point.

Recently, the cold dark matter model has been severely challenged by the observations, perhaps fatally. In 1990, S. J. Maddox, G. Efstathiou, W. Sutherland, and J. Loveday of Oxford University compiled the 2-dimensional positions (without depth or redshift data) of 2.5 million galaxies in the southern sky—the largest survey of galaxies ever done—and claim that they see too much clumpiness of galaxies on large scales to be explained by the cold dark matter model. In January of 1991, a collaboration of scientists from Queen Mary and Westfield College, University of Durham, University of Oxford, and University of Toronto analyzed the 3-dimensional locations of about 2,000 galaxies distributed across the whole sky. This galaxy survey, which was unique in combining redshift information with large sky coverage, strongly suggests that there is more clustering of galaxies on scales larger than 30 million light years than can be explained by the cold dark matter model. These new observations of substantial inhomogeneities on large scales confirm earlier work in 1983 by Neta Bahcall of Prince-

The distribution of 2 million galaxies in an area covering 10 percent of the sky in the southern hemisphere, recently compiled by Steve Maddox and collaborators at Oxford University. The 2-dimensional positions of galaxies were recorded by an automated computer-controlled scanning machine. (Redshift data, and distances, are not compiled for these galaxies.) Each white dot indicates more than 20 galaxies; each grey dot indicates between 1 and 19 galaxies. Black squares are artifacts of the recording procedure; other black areas are true absences of galaxies. The small bright patches are individual galaxy clusters. The larger elongated bright areas are superclusters and filaments. The overall mottled pattern is caused by small-scale clustering of galaxies.

ton University and Raymond Soneira of AT&T Bell Laboratories. Bahcall and Soneira found that on scales of several hundred million light years, clusters of galaxies themselves show more clustering than can be easily explained by the cold dark matter mode. The Great Attractor, discovered several years later, also implies mass inhomogeneities on larger scales than can be easily explained by the cold dark matter model. In consideration of all of these observations, most astronomers believe the cold dark matter model is in serious trouble.

Some scientists believe that the observed inhomogeneities, and especially the sharpness of those inhomogeneities, require either other physical forces beside gravity or special initial conditions to explain what we see in our local region of the universe. For example, in 1981 Jeremiah Ostriker of Princeton and Lennox Cowie of the University of Hawaii proposed that gas pressure, generated by the explosions of stars, may have been the major force in forming galaxies and groups of galaxies. A similar idea was proposed much earlier by Doroshkevich, Zel'dovich, and Novikov of the Soviet Union. Such pressure waves might travel outward from various centers of explosions, expelling all the gas from a spherical cavity and then depositing it at the edge, where galaxies could then form. But even this nongravitational explanation for the observed distribution of galaxies apparently cannot explain inhomogeneities on scales as large as 30 million light years and larger.

Finally, there is the cosmic background radiation. Whether or not the matter of the universe is sprinkled about evenly when viewed on scales of billions of light years, it is certainly uneven and structured on smaller scales. If the observed structures have grown from small lumps in the distant past, as required by both the pancake model and the gravitational hierarchy model, then those initial lumps must have produced some unevenness in the cosmic background radiation. Indeed, such unevenness is demanded by all current theories of the formation of cosmic structures.

So far, no unevenness has been observed. From the measurements of the Cosmic Background Explorer and from other new measurements done by several groups in the United States, astronomers have recently determined that any variations in the intensity of the cosmic background radiation must be less than several parts in 100,000. Older theories of galaxy formation have been demolished. Luckily for theorists, new theories, which require large amounts of missing mass, predict much less unevenness in the cosmic background radiation. The new theories are still safe. In the next decade, detectors now being developed will have the required sensitivity to challenge the new theories: variations at 1 part in a million. If no unevenness in the cosmic background radiation is found at these increased sensitivities, then there will be a serious problem. (Very new observations hint at positive detections of unevenness at the level of one part in 100,000—which would provoke a great sigh of relief among theorists—but these observations have not been fully analyzed.) In recent years, astronomers have become increasingly worried about reconciling the smoothness of the cosmic background radiation with the lumpiness of matter nearby.

The large-scale structure of the universe is the most active area in cosmological research today. Given all the difficulties mentioned, most cosmologists feel that *no* current models are satisfactory. The underlying big bang model itself could be in jeopardy. Whether the current models will or will not prove adequate, for the first time, theory and observation of large-scale structure are within arms length of confronting each other. Both theoretically and observationally, an understanding of the large-scale structure of the universe is at the top of most people's list of the outstanding problems in cosmology.

Granted the bubbles and walls and strings of galaxies, we must not forget that the universe is still remarkably smooth compared with what it might be. The density of galaxies and the rate of expansion of the universe are approximately the same in every direction. And the intensity of the incoming cosmic background radiation varies by less than a part in 10,000 as our radio telescopes sweep the celestial sphere. Cosmologists certainly have to explain why galaxies cluster as they do, but cosmologists must also explain why the big picture is so smooth.

CHAPTER • EIGHT

Instruments and Technology

We have said little about the importance of instruments and hardware in astronomy, but the dramatic progress of the 1980s in mapping the locations and motions of galaxies was fueled in large part by technology. New electronic light gathering and recording devices and computers have allowed images and colors of galaxies to be recorded faster, digitized, and processed in automated procedures. Indeed, much of the progress in all of astronomy has been driven by instruments and technology.

To illustrate the technological revolution in astronomy, we will first consider optical astronomy. Most astronomical objects emit light at many different wavelengths, with wavelengths both smaller and larger than those the human eye can detect. Optical light is light we can see.

There are two basic steps in detecting a light source: the light must be collected, and it must be recorded. In our own eye, the pupil collects the light, and the retina records it. Telescopes were invented to collect more light than can the human eye. One of the first big telescopes in the United States was the 100-inch telescope at Mt. Wilson, California, built around 1920 and used by Edwin Hubble and others. The "100-inches" refers to the diameter of its opening. With such a large pupil, a telescope can gather much more light and thus see much fainter objects than can a human eye. In

Moonlight view of the 200-inch Hale Telescope on Mt. Palomar.

particular, every doubling of the diameter of a telescope's opening quadruples the amount of light it receives and thus allows it to see objects 4 times as faint. As of 1991, the largest telescope in the United States is the 200-inch at Mt. Palomar, California. It was completed in 1949. In the early 1990s, astronomers will complete a number of new telescopes with diameters ranging from 300 inches to 400 inches. The first of these new large telescopes, the 396-inch (10-meter) Keck telescope, will be located in Hawaii. In the spring

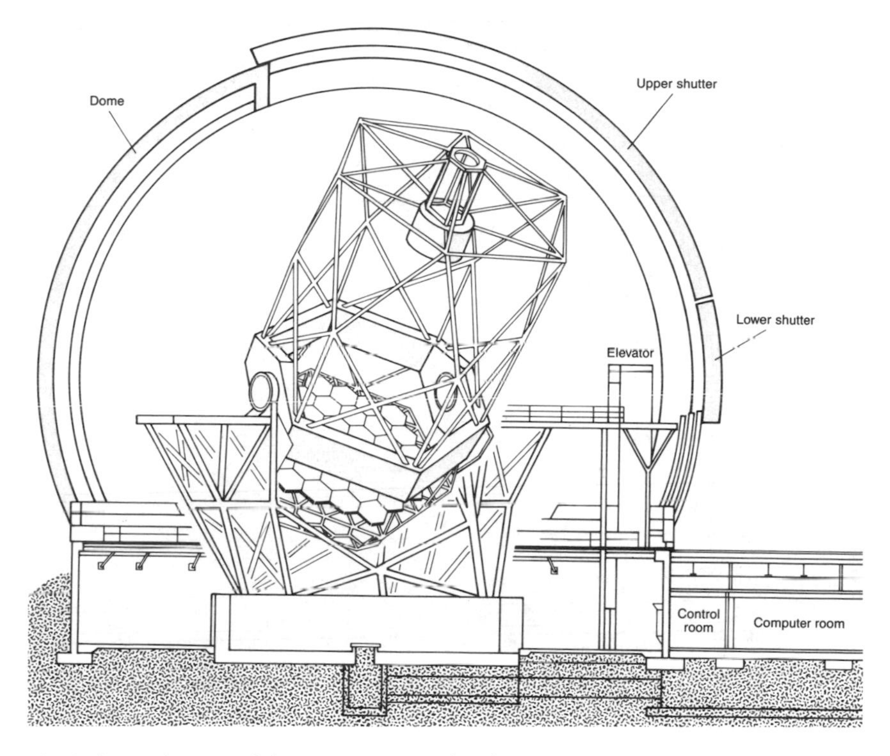

Artist's rendering of the 10-meter Keck telescope, on Mauna Kea in Hawaii.

of 1990, the Hubble Space Telescope was placed in orbit about the earth, where it is immune to the distorting and smearing effects of the atmosphere. The Hubble Space Telescope, unfortunately, has a flaw in its mirrors. When it is fixed, the Hubble Space Telescope should be able to make out details as small as 0.00003 degrees in angle, equivalent to the width of a penny 25 miles away.

Until recently, light from astronomical objects was recorded by a photographic plate placed at the "back end" of the telescope. The grains in a photographic emulsion, however, respond to only about 1 percent of the incoming light. To record the weak light from a distant galaxy, long exposure times were needed, making it difficult to undertake large surveys of many galaxies. In the early 1960s, telescopes were fitted with the first electronic imaging device, the

The Hubble Space Telescope, launched in the spring of 1990.

image tube. The image tube sits above the photographic plate and greatly amplifies the incoming light. The photographic plate still records the light, but so much more light comes in that the needed exposure time is about 10 times less. The image tube was the first big advance in technology.

The next leap forward came in the period 1970–1972, when the photographic plate was replaced by a computer-driven digital detector. Such devices were developed independently by John B. Oke of the California Institute of Technology, Alexander Boksenberg of the Greenwich Royal Observatory in England, and Joseph Wampler and Lloyd Robinson at the Lick Observatory of the University of California at Santa Cruz. A digital detector translates incoming light into electrical signals, rather than darkened grains on a photographic plate. The electrical signals can be digitized and stored in a computer, and the stored data can be easily manipulated.

The "z-machine," a digital detector designed by Stephen Shectman and used in the Center for Astrophysics redshift survey to record galaxy redshifts in an automated process.

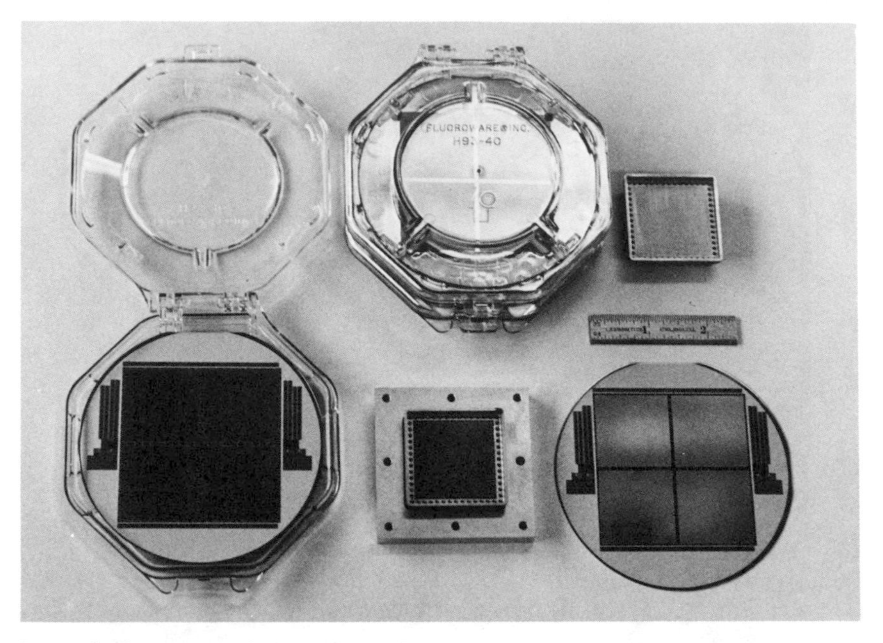

Several charge-coupled devices (CCD), which replaced both the photographic plate and the image tube.

For example, if it is known that the image of a distant galaxy was partly obscured by the light from a foreground star, the computer can electronically subtract out the unwanted light and reconstruct a clear picture of the galaxy. Digital detectors were made possible, in part, by the revolution in computer microchip technology of the late 1960s. In the mid to late 1970s, Stephen Shectman of the Mt. Wilson and Las Campanas Observatories designed a cheap and simple digital detector that was reproduced in a number of observatories. Sometimes called the z-machine or Shectograph, Shectman's device has been used in the Center for Astrophysics redshift survey, for example.

A more recent high-technology device to transform optical astronomy is the charge-coupled device, or CCD. The CCD is a digital detector with such high efficiency and sensitivity that an image tube is not needed. Indeed, charge-coupled devices represent an additional improvement of a factor of 5 or 10 over image tubes

James Gunn was born in Livingstone, Texas, in 1938, educated at Rice University and the California Institute of Technology, and is now professor of astrophysics at Princeton. Like Dicke, Gunn is outstanding both as an observer and as a theorist. He has designed and built highly sensitive light-recording devices and, with others, predicted a relationship between the possible types of subatomic particles and the cosmic helium abundance.

in exposure time needed. The CCD replaces both the old photographic plate and the image tube. First developed in 1969 at Bell Laboratories, CCDs were introduced into astronomy largely through the work of James Gunn of Princeton and James Westphal of the California Institute of Technology.

Another recent high-technology device is called fiber optics, first used in 1985. Here, the light from the telescope is channeled into many different glass fibers, each one carrying the detected light from a single object, like a galaxy, and conveying that light to a separate device for analysis. In this way, the colors, and redshifts, of many different galaxies may be recorded simultaneously. Each glass fiber is a long, thin tube about one hundredth of an inch in diameter. There are typically about 100 glass fibers attached to a telescope, allowing about 100 different galaxies to be analyzed at once.

The latest electronic innovation, still under development, is called "adaptive optics." Adaptive optics is an electronic feedback mechanism capable of correcting for the distorting effects of the earth's atmosphere and thus allowing much sharper images of astronomical objects. The earth's atmosphere is constantly shimmering, because of moving pockets of air and changes in temperature, and such shimmering causes passing light rays to bend one way and then another. In effect, the shifting atmosphere acts as a rapidly changing lens, smearing out and defocusing images. In adaptive optics, motorized cushions are placed behind the telescope's secondary mirror and constantly reshape the mirror's surface to counteract the defocusing effect of the atmosphere. The cushions are given their instructions by a computer, which analyzes the image of a "guide star" in the same field of view as whatever the telescope is looking out. With no atmospheric distortion, the image of a star should be a single point of light. By analyzing how the actual image of the guide star differs from a point, the computer can infer the distortion of the atmosphere and tell the cushions how to alter the mirror to bring the guide star, and all objects near it, back into sharp focus. Corrections must be made rapidly because the atmosphere is rapidly shifting. In practice, the computer will reanalyze the image of a guide star and give new instructions to the reshaping cushions every 0.01 to 0.1 seconds.

An important consequence of the new high-technology instruments is that big observational programs once requiring large telescopes can be now carried out with moderate-sized ones. Given the heavy demand for large telescopes, the possibility of doing a lengthy project on a more available, smaller telescope can mean the difference between doing or not doing the project.

So far, we have considered only instruments that record optical light, which is the only light that astronomers could detect for thousands of years. However, a number of new kinds of telescopes and instruments can detect radiation that is invisible to the human eye. Indeed, optical light makes up just a small fraction of the electromagnetic spectrum. Infrared radiation has longer wavelengths than optical light, and radio waves have still longer wavelengths. On the other side of the spectrum, with wavelengths shorter than visible (optical) light, come ultraviolet radiation, X-rays, and gamma rays, in a succession of ever shorter wavelengths. Although these various radiations have different names, they are all similar forms of energy differing only in their wavelengths. For the visible region of the electromagnetic spectrum, the different wavelengths correspond to different colors. Blue light has the shortest wavelength of visible light, and red the longest.

In the 1930s new communication devices led to the reception of radio waves from space. Radio waves were the first form of invis-

Infrared photograph of the Milky Way, obtained by the Diffuse Infrared Background Experiment aboard the Cosmic Background Explorer satellite. The Earth lies in the plane of the disk and is part of that disk.

ible radiation to be detected by human instruments. Now, radio dishes are common. The world's largest single-unit radio telescope is the Arecibo Observatory in Puerto Rico. This telescope is a giant dish 1,000 feet in diameter, made of perforated aluminum panels that focus incoming radio waves. Unfortunately, radio waves and optical light are the only radiations that pass freely through the earth's atmosphere, without absorption (although limited wavelengths of other radiations do so as well). To detect other forms of electromagnetic radiation, instruments must be launched above the earth's atmosphere.

Since the 1940s, a series of rockets and satellites have recorded infrared radiation, ultraviolet radiation, and X-rays from space. Such radiations, all invisible to the eye, have revealed completely new features of many astronomical objects and announced some objects not before known. In addition to the logistical problem of getting above the atmosphere, each different kind of radiation requires a different kind of instrument to detect it.

An example of the new space-based astronomical instruments was the Einstein X-ray Observatory, which orbited the earth from 1978 to 1981. This instrument had specially designed mirrors that could focus incoming X-rays and form images of astronomical objects in "X-ray light." Planned for the 1990s is a successor to the Einstein Observatory called the Advanced X-ray Astrophysics Fa-

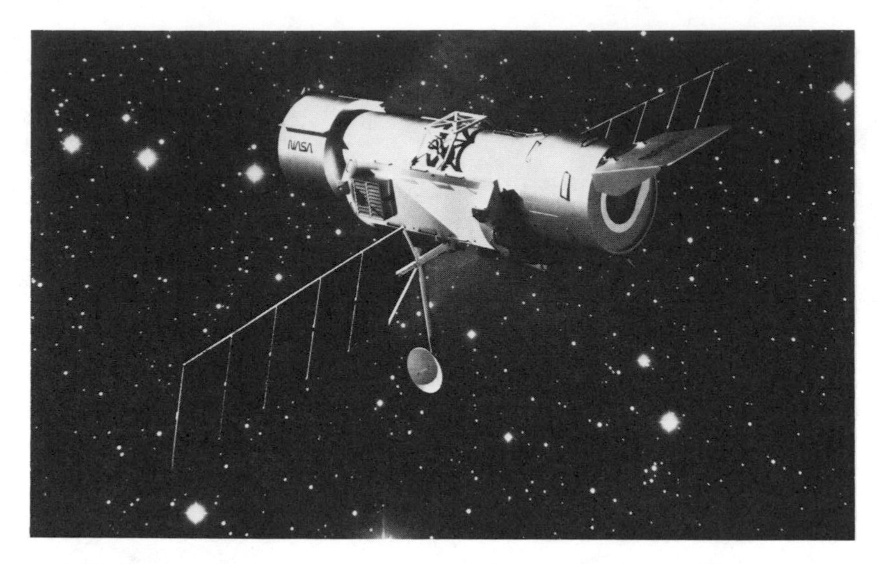

Artist's rendering of the proposed Advanced X-Ray Astrophysics Facility (AXAF).

Artist's rendering of the proposed Space Infrared Telescope Facility (SIRTF).

cility, or AXAF for short. AXAF will also be able to focus X-rays, and it will have 10 times the angular resolving power and 100 times the sensitivity of its predecessor. Another recent space-based instrument was the Infrared Astronomical Satellite, in earth orbit in the mid 1980s. Many types of molecules in space reveal their identity by their infrared emission. In the 1990s, astronomers hope to launch a more advanced infrared satellite called the Space Infrared Telescope Facility, or SIRTF for short. SIRTF will have 10 times the angular resolution of its predecessor and *several thousand to one million* times the sensitivity.

Miniaturized computers onboard these orbiting satellites control the program of observations of the detectors, directing them to point in different directions at different times. After an orbiting telescope has taken its images, computers and other high-technology devices relay those images to the ground. There are no film drops. The pictures and data from the telescope are recorded in the form of electrical impulses, digitized (converted into bits of information represented by patterns of zeros and ones), and then transmitted by radio to earth.

C H A P T E R • N I N E

Initial Conditions and Quantum Cosmology

INITIAL conditions play a peculiar role in cosmology. In general, the initial conditions and the laws of nature are the two essential parts of any physical calculation. The initial conditions tell how the forces and particles are arranged at the beginning of an experiment. The laws tell what happens next. For example, the motions of balls on a pool table depend both on the laws of mechanics and on the initial positions and speeds of the balls. Although such initial conditions must be specified at the beginning of an experiment, they can also be calculated from previous events. In the case of the pool balls, the initial arrangement was the result of a previous arrangement, which was ultimately a result of how the first ball was struck with the cue. Thus, the initial conditions of one experiment are the final conditions of a previous one. This notion fails for the initial conditions of the universe. By definition, nothing existed prior to the beginning of the universe, if the universe indeed had a beginning, so its initial conditions may have to be accepted as an incalculable starting point—the particular arrangement of balls in the rack before the first break. Such an incalculable starting point distresses physicists, who want to know *why*.

The flatness and horizon problems are especially compelling if we believe that the universe could have begun with many possible initial conditions and physical processes, only a small fraction of which would have led to a universe as homogeneous and nearly flat as our own. It is certainly possible to postulate that the universe began with a uniform density and temperature and began with a near perfect balance between gravitational energy and kinetic energy of expansion. The question is whether such initial conditions are plausible. Are they probable or improbable? Probability arguments traditionally require that an experiment be carried out in a large number of identical systems or that an experiment be repeated many times on a single system. For example, you can sensibly speak about the probability of a car coming by your house between 8:00 a.m. and 8:01 a.m. on a Tuesday morning if you have looked out your window every Tuesday morning for a thousand Tuesdays and compiled some statistics. A thousand universes are not available.

How might the initial conditions of the universe have been determined? Did the universe suddenly appear at $t = 0$? The standard big bang model, based on Einstein's theory of gravity, requires that the universe exploded into existence from a state of infinite density. However, scientists agree that this model is not complete at extremely high densities of matter. In particular, Einstein's theory of gravity does not incorporate the physics of quantum mechanics. All other modern theories in physics do. In the 1920s physicists discovered that all phenomena of nature have a dual particle-like and wave-like behavior. In some cases, an electron acts like a particle, occupying only one position in space at a time, and in other cases it acts like a wave, occupying several places at the same time. The theory of this strange behavior is called quantum mechanics. The wave-particle duality of matter leads to an intrinsic uncertainty in nature, that is, an uncertainty not arising from our ignorance or inability to measure but an absolute uncertainty. Nature must be described by probabilities, not by certainties.

Physicists have tried and so far failed to find a complete theory of gravity that includes quantum mechanics. When calculations are done in all proposed theories of "quantum gravity," those calculations lead to infinities. Physicists are not sure whether the problem

Stephen Hawking was born in Oxford, England, in 1942, educated at Oxford University and Cambridge University, and is now professor of mathematics at Cambridge. Hawking's contributions to cosmology include work with Roger Penrose on the cosmological singularity theorems, showing the likelihood of the big-bang beginning of the universe, and more recent work attempting to formulate the initial conditions of the universe. Since the 1960s, Hawking has been afflicted with an incurable and degenerative neuromuscular disease.

is technical or conceptual. Even without a complete theory of quantum gravity, however, it can be estimated that quantum mechanical effects would have been crucial during the first 10^{-43} seconds after the beginning of the universe, when the universe had a density of 10^{93} grams per cubic centimeter and larger. (Solid lead has a density of about 10 grams per cubic centimeter.) This period is called the quantum era or Planck era, and its study is called quantum cosmology. Since the entire universe would have been subject to large uncertainties and fluctuations during the quantum era, with matter and energy appearing and disappearing in large quantity out of a vacuum, the concept of a beginning of the universe might not have a well-defined meaning. However, the density of the universe during this period was certainly huge beyond comprehension. For all practical purposes, the quantum era could be considered the initial state, or beginning, of the universe. Correspondingly, whatever quantum processes occurred during this period determined the initial conditions of the universe.

Using the general ideas of quantum mechanics, but without a detailed theory of quantum gravity, Stephen Hawking of Cambridge University, James Hartle of the University of California at Santa Barbara, and others have recently attempted to *calculate* the expected initial conditions of our universe. Such calculations are very different from observing what the universe is today and working backward to figure out what the universe was like near its beginning. Hawking and Hartle are proposing to calculate how the universe *had* to be created—consistent with the general concepts in quantum theory and relativity theory—and then to work forward from there. The details of such a calculation must await a theory of quantum gravity; even then, the calculation may be too difficult to carry out in practice. Yet if such a calculation could be reliably done, the initial conditions would not have to be taken as a given. Initial conditions would be on the same footing as the laws of nature. All aspects of the universe could in principle be calculated and explained.

For some time, many scientists thought that the notion of an ultra-high-density beginning of the universe was an artifact of the idealized assumptions of the big bang model, such as the assumption of homogeneity. However, in the mid 1960s Roger Penrose and Stephen Hawking mathematically proved that even if the universe is not homogeneous at all, its current expansive behavior, together with the theory of general relativity, require that the universe had to have been enormously denser in the past—at least as far back in time as classical general relativity applies, that is, to the Planck era. Thus, it seems that quantum cosmology must eventually be dealt with to understand the initial state of the universe.

Some cosmologists, especially the theorists, believe that a real understanding of why the universe is as it is will elude us until we understand the initial conditions of the universe and have a reliable theory of quantum gravity. Such a theory could be many years away.

C H A P T E R • T E N

Particle Physics, the New Cosmology, and the Inflationary Universe Model

IN THE 1970S an important change occurred in theoretical cosmology. A group of physicists with expertise in the theory of subatomic particles joined astronomers to work on cosmology. They brought a fresh stock of ideas and a new set of intellectual tools to bear on the question of *why* the universe has the properties it does, not just *what* those properties are.

In the "old cosmology," before the 1970s, most cosmologists concerned themselves with measuring the distances and motions of galaxies, the formation and composition of galaxies, the rate of expansion of the universe, and the average density of matter. In the "new cosmology," scientists have seriously begun to ask such questions as why matter should exist at all. Where did it come from? Why is the gravitational energy of the universe so nearly equal to its kinetic energy of expansion (the flatness problem)? Why does the cosmic background radiation, arriving from billions of light years away, appear precisely the same regardless what direction the telescope is pointed in (the horizon problem)? Why is the ratio of photons to baryons in the universe a billion to one, rather than some much bigger or smaller number? Why did the universe begin in such a high degree of orderliness (the entropy problem)? The question of *why* was added to *what* and *how*. Some of these questions had been posed before, by a handful of scientists, but they had been largely dismissed or abandoned because no one had good ideas about solving them. Many scientists had regarded such questions as lying outside the purview of science. Particle physics expanded the science of cosmology.

A prelude to the future collaboration between particle physics and cosmology could be heard in the 1960s. It concerned the number of different types of elementary particles. Theories of elementary particles and forces depend crucially on how many types of elementary particles there are, just as Aristotle's theory of the universe depended on his five elements: fire, water, air, earth, and ether. According to theoretical calculations first done in the 1960s, the amount of helium produced in the nuclear reactions of the early universe should have depended on the number of types of certain subatomic particles called leptons. (The electron, for example, is one type of lepton.) The more types of such particles, the more helium should have been produced. Thus, from the actual abundance of helium, which is measured to be about 25 percent, we can determine the number of types of leptons.

The expected relationship between helium abundance and lepton types was first suggested by Hoyle and Tayler in 1964 and later independently suggested by Robert Wagoner of Stanford in 1967. V. F. Shvartsman of Moscow State University did the first quantitative calculation of the effect in 1969. In 1977 Gary Steigman of the Bartol Research Foundation, David Schramm of the University of Chicago, and James Gunn of Princeton University, without knowledge of previous work, rediscovered the effect and did a more detailed calculation. At the present time, three types of leptons are known—electrons, muons, and taus, and their associated antiparticles and neutrinos—but some theories of particle physics predict that there could be many more. The theoretical calculations of Schramm and coworkers indicated that there could be *at most* one new type of lepton, for a maximum total of four. Otherwise, the fraction of helium produced in the early universe, as calculated by the big bang theory, would disagree with the fraction observed. Experiments in 1989 to probe the number of types of leptons,

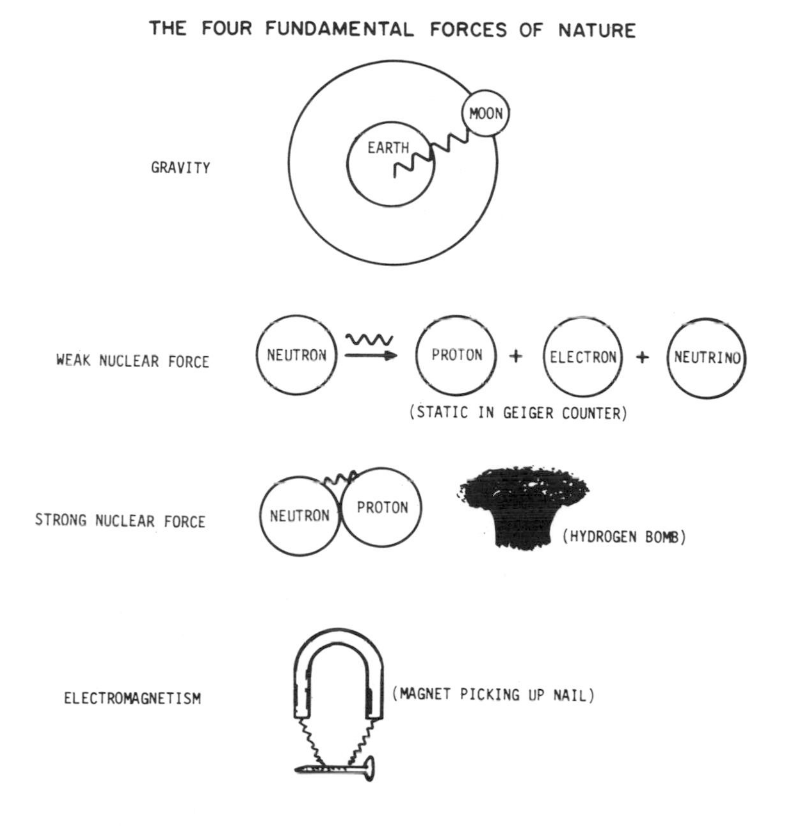

carried out in the giant CERN particle accelerator in Geneva and at the Stanford Linear Accelerator in California (SLAC), indicate that there are *no* new types of leptons. The three so far observed are all there are. This confirmation of a result predicted from cosmology, using particle-physics technology, delights physicists and astronomers alike.

Most of the physicists involved in the above calculations had a background in astronomy and cosmology. However, in the mid 1970s particle physicists with barely any roots in cosmology began venturing into the field. A principal motive was to test the new grand unified theories of physics. Grand unified theories, often called GUTs, propose that three of the four fundamental forces of

Steven Weinberg was born in New York City in 1933, educated at Cornell University and Princeton University, and is now professor of science at the University of Texas at Austin. Weinberg was one of the first physicists to apply new theories of elementary particles to cosmology. He and others pointed out that effects of grand unified theories, occurring in the early universe, could explain the ratio of photons to baryons. Weinberg's popular book, *The First Three Minutes* (1977), introduced much of the public and many scientists to cosmology. In 1979 Weinberg won the Nobel Prize in physics for his theoretical work on unifying the electromagnetic and weak nuclear forces.

nature are actually different versions of a single, underlying force. (In a similar way, electricity and magnetism are not really two separate forces, because they can generate each other. A magnet moving inside a coil of wire produces an electrical current, and an electrical current flowing through a wire produces a magnetic field wrapping around the wire.) The three fundamental forces combined in grand unified theories are the electromagnetic force; the strong nuclear force, which holds together the subatomic particles in the atomic nucleus; and the weak nuclear force, which is responsible for certain kinds of radioactivity. Gravity, the fourth force, has not been successfully unified because of the difficulties in combining gravity with quantum mechanics.

A grand unified theory has long been the holy grail of physicists. Since ancient times, physicists have sought minimalist explanations of nature. Theories with four basic particles are considered better than theories with ten. One force that explains the fall of apples and the orbit of the moon is better than two. There is still little experimental evidence that grand unified theories are correct. One practical difficulty of testing them is that their effects become significant only at extremely high temperatures, much higher than can be created on earth or even at the centers of stars. Such extreme temperatures could be achieved at only one place, or rather, at only one time—in the infant universe, when all the matter of the universe was in the form of an ultra-high-temperature gas of subatomic particles. Thus, particle physicists became interested in cosmology.

Grand unified theories represent a bold leap beyond familiar terrain. The highest energies well tested in particle accelerators on earth correspond to a temperature of about 10^{14} degrees centigrade. Grand unified theories concern temperatures of about 10^{28} degrees, or about 100 trillion times higher. Put in other terms, the universe was about 10^{-8} seconds old when its material was at a temperature of 10^{14} degrees. Any further extrapolation back in time toward the big bang, toward higher temperatures, enters the realm of speculation. Yet, cosmologists have been forced to speculate. Many of the properties of the universe may have been determined in the first 10^{-8} seconds and much earlier.

If grand unified theories are correct, then their most interesting effects would have happened when the universe was about 10^{-35} seconds old. This is the epoch when the cosmic temperature was 10^{28} degrees and when the various combined forces would have begun disengaging from one another. One of the most speculative and spectacular phenomena predicted by GUTs during this epoch is the formation of extremely thin strands of energy called cosmic strings. According to the theory, a cosmic string has a diameter of about 10^{-30} centimeters and the energy equivalent of 10^{26} tons of TNT per inch of its length. Cosmic strings, if they exist, might have seeded the formation of galaxies and clusters of galaxies by virtue of their intense energy concentration and resulting gravitational effects. The universe today could be populated by cosmic strings of various lengths, ranging from fractions of inches to billions of light years. As of yet, there are no reliable theoretical estimates of how many such strings should exist and no observational evidence that they exist at all.

In 1978 and 1979 several groups of particle physicists, including Steven Weinberg, then at Harvard and now at the University of Texas, pointed out that processes resulting from the grand unified theories, acting about 10^{-35} seconds after the big bang (well after the highly uncertain quantum era of the first 10^{-43} seconds), could

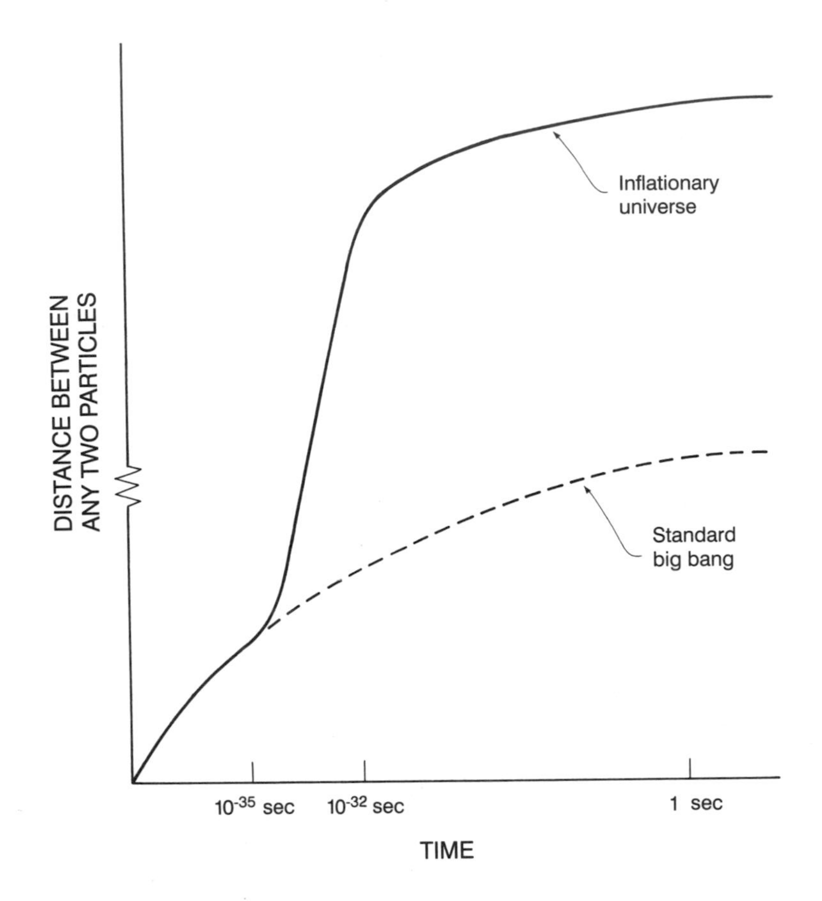

The expansion of the universe in the standard big bang model and in the inflationary universe model.

explain why the ratio of photons to baryons had the value it did. Although there was a fair amount of ambiguity in these calculations, arising in part out of an uncertainty about which grand unified theory was most likely correct, cosmologists were enormously excited that such a previously mysterious cosmological quantity could be calculated at all. Weinberg was a highly respected theoretical physicist in the area of subatomic particles and was soon to win a Nobel Prize for his earlier work in particle physics. Encouraged by Weinberg's example, a new generation of young

Alan Guth was born in New Brunswick, New Jersey, in 1947, educated at the Massachusetts Institute of Technology, and is now a professor of physics there. Trained as an elementary particle physicist, Guth was one of the principal architects of the influential inflationary universe model, a modification of the big bang model.

particle physicists decided that they too would work on cosmology.

One of the young physicists inspired by Weinberg was Alan Guth. In late 1978 Guth learned about the flatness problem from a lecture at Cornell given by Robert Dicke. About a year later, while a postdoctoral fellow at the Stanford Linear Accelerator, Guth proposed a modification to the big bang model that provided a natural explanation of the horizon and flatness problems. Guth's new cosmological model is called the inflationary universe model and has caused a major change in cosmological thinking. Some of the ingredients of the inflationary universe model had been previously discussed by others, including R. Brout, F. Englert, and P. Spindel of Belgium; Demosthenes Kazanas of the NASA Goddard Space Flight Center in Maryland; Martin Einhorn of the University of Michigan and Katsuhiko Sato of Japan; and A. A. Starobinsky of the Landau Institute in Moscow. But it was Guth's clear statement of the model and its assets that galvanized the scientific community.

The essential feature of the inflationary universe model is that, shortly after the big bang, the infant universe went through a brief and extremely rapid expansion, after which it returned to the more leisurely rate of expansion of the standard big bang model. By the time the universe was a tiny fraction (perhaps 10^{-32}) of a second old, the period of rapid expansion, or inflation, was over. Under some conditions, the inflationary behavior of the infant universe is a natural consequence of grand unified theories. These theories predict that at the moment when the single unified force began acting as separate forces, the energy and mass of the universe existed in a peculiar state called a false vacuum, which behaves as if it had negative gravity. Negative gravity repels, so that instead of retarding the rate of expansion, it speeded up the rate of expansion. The inflationary period would have ended when the energy and mass of the universe changed from the peculiar state back to the normal state, with attractive gravity.

The epoch of rapid expansion could have taken a patch of space so tiny that it had already homogenized and quickly stretched it to a size larger than today's entire observable universe. Quantitative estimates can be made, although these estimates are uncertain due to ignorance about the details of the underlying grand unified theory and the resulting ignorance about exactly when the inflationary epoch began and ended. For purposes of illustration, we will assume that the inflationary epoch began when the universe was 10^{-35} seconds old and ended when it was 10^{-32} seconds old. At the beginning of the inflationary epoch, the largest region of space that could have homogenized would have been about 10^{-35} light seconds in size, or about 10^{-25} centimeters, much smaller than the nucleus of an atom. At the end of the inflationary epoch, this tiny homogenized region would have been stretched to something like 10^{400} light years. For comparison, at the end of the inflationary period the region of space from which we now detect the cosmic background radiation was only about 1,000 centimeters in size, as was today's entire observable universe. Thus, the inflationary expansion would have homogenized the universe over an extremely large region, far larger than any region from which we have data. The horizon problem is solved. Regions of space that appear to have never been close enough to have exchanged heat, according to extrapolations into the past based on the standard big bang model, were actually much closer, based on the inflationary universe model.

The inflationary universe model also solves the flatness problem. Regardless of the initial curvature of the universe—whether curved

Andrei Linde was born in Moscow in 1948, educated at Moscow State University and at the Lebedev Physical Institute, and is now professor of physics at the Lebedev. With Guth, Linde was one of the creators of the inflationary universe model. Linde has proposed a new version of that model, called chaotic inflation, in which the universe continually and randomly spawns new universes. Although Linde's work is very mathematical, he describes himself as more intuitive than technical.

in the manner of an open universe or curved in the manner of a closed universe—any observable patch of the universe would be very nearly flat after the period of rapid expansion, just as a dime-size area of the surface of a beach ball would appear nearly flat after the ball has been inflated to a mile in diameter. The inflationary universe model firmly predicts that the universe today should be extremely close to flat. Equivalently, the average density of matter should be extremely close to its critical value—much closer than the one-tenth the critical value estimated from current observations. On this basis, the model can be either supported or ruled out from observational evidence. If the hypothesized missing mass is found to exist, the model is supported; if it can be shown that the missing mass does not exist, the model is ruled out.

The inflationary universe model explains some of the outstanding problems of cosmology without resorting to the explanation-by-initial-conditions argument. In fact, it provides a physical mechanism by which the initial conditions of the universe may have been *irrelevant*—a notion that pleases and relieves many physicists. And the theory is elegant, combining subatomic and astronomical physics as it does. The inflationary universe model also has led to a new picture of the cosmos. Since the universe was enormously stretched out during the inflationary epoch, it may be vastly larger than we thought. Consequently, what we see from looking out of our largest telescopes—the small patch of space that we call the observable universe—may tell us very little about the universe as a whole. In this sense, the inflationary universe model has made us even smaller than we were before.

Andrei Linde of the Lebedev Physical Institute in Moscow has taken this idea further in what he calls chaotic inflationary universe models. Linde speculates that under certain conditions an inflating universe can separate into different pieces, completely cut off from each other, in effect different universes. Each of the new pieces can repeat the process, in a random way, with each universe spawning many new universes. Individual universes would come and go, like Anaximander's worlds, but the collection of universes would live forever. Some of the new universes might have very different properties from ours—different forces, different types of particles, even different dimensionalities of space. In such a scenario, it would be impossible for us ever to learn about more than a tiny fraction of the possibilities and realities of nature.

The original inflationary universe model, as proposed by Guth in early 1980, had difficulties. In Guth's model, the special kind of matter and energy with negative gravity did not smoothly fill the universe but was scattered here and there. As a result, some parts of the universe expanded rapidly while others expanded slowly. The rapidly expanding pieces became more and more homogeneous, but the universe as a whole was left with a highly riddled structure, in conflict with the observations. This difficulty was solved in 1982 by a new version of the inflationary universe model developed by Paul Steinhardt and Andreas Albrecht at the University of Pennsylvania and independently by Andrei Linde. In this revised model, all parts of the universe began inflating and stopped inflating at the same time. Approximate homogeneity was maintained. Unfortunately, the new inflationary universe model of Linde, Albrecht, and Steinhardt also has serious problems. Theoretical calculations suggest that the model, while much more homogencous than the original inflationary model, predicts inhomogeneities in the early universe that are sufficiently large to spoil the observed smoothness of the cosmic background radiation. Moreover, a somewhat special kind of grand unified theory is needed to lead to a suitable phase of rapid expansion.

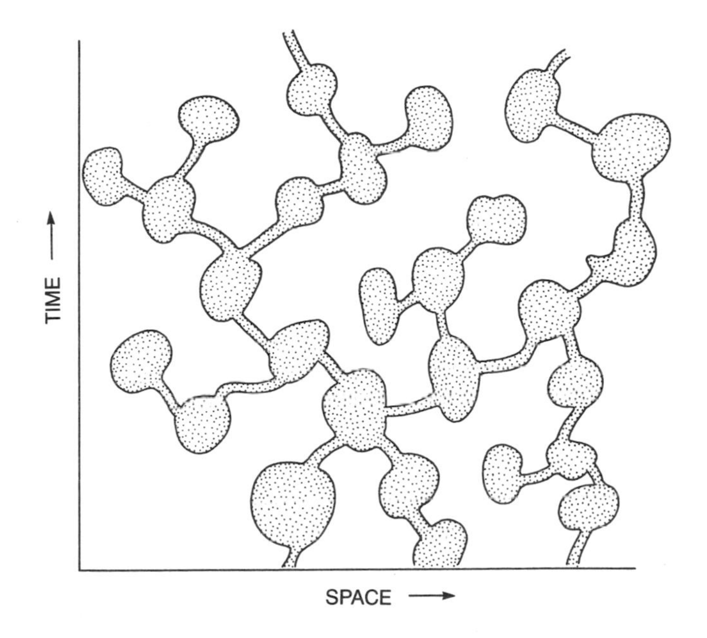

An illustration of the chaotic inflationary universe model. Each bubble is essentially a separate universe and, in time, spawns new universes.

Furthermore, there is no direct observational evidence to confirm the inflationary universe model. In fact, one of the critical predictions of the model is inconsistent with observations at the present time. Almost any version of the inflationary universe model predicts that the value of omega today should be extremely close to 1, even 10 billion years after the big bang. But the actual measured value of omega is about 0.1, although this value is somewhat uncertain. In other words, we have detected about one tenth as much cosmic mass as is required by the inflationary universe model. Scientists who believe on theoretical grounds that the model is right must therefore have faith that an enormous amount of mass is hiding from us, escaping detection, perhaps in a uniform and tenuous dark gas of particles between galaxies. In addition, the missing mass required by the inflationary universe model cannot be composed of the ordinary matter that makes up atoms. Agree-

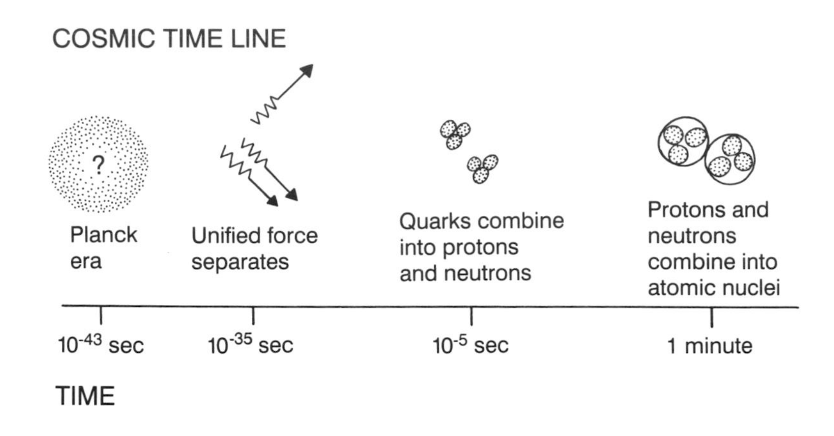

ment between the observed abundance of helium and the theoretical calculations of its production in the early universe requires that the omega derived from ordinary matter, such as protons and neutrons, cannot be larger than 0.1. The missing mass must therefore consist of some exotic species of matter.

Despite all these difficulties, the general features and results of the inflationary universe model are so appealing that most cosmol-

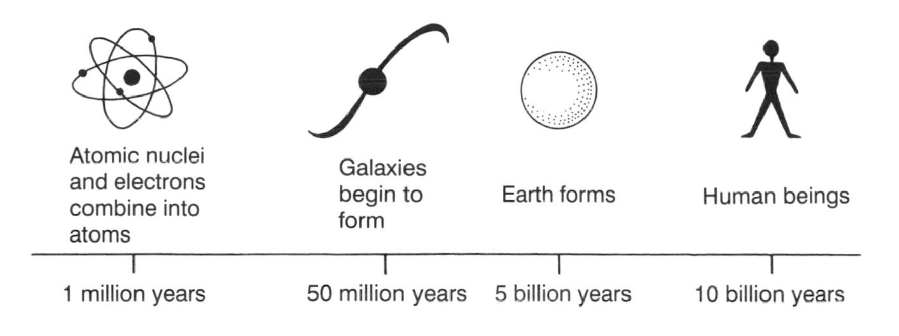

ogists believe some form of the idea is correct, even if it may differ in detail and mechanism from the original proposals of Guth, Linde, Albrecht, and Steinhardt. However, the inflationary model may not offer a good solution to Penrose's entropy problem. Some physicists, therefore, believe that the model will be replaced by a more fundamental understanding of the origin of the universe.

CHAPTER • ELEVEN

The Anthropic Principle

FOR THE last couple of decades, a small but eminent group of physicists have addressed the problem of initial conditions in cosmology in terms of the conditions needed to create life. It seems plausible that not any conceivable physical conditions would allow life to form; the fact of our existence, therefore, might limit what possibilities should be considered. This notion is called the anthropic principle.

There are two forms of the anthropic principle, the weak and the strong. The weak form states that life can arise and exist only during a certain epoch of our universe. The strong form states that only for a special kind of universe could life arise at all, at any epoch. The weak anthropic principle restricts itself to the universe we live in; the strong implicitly refers to many possible universes. Some cosmologists have not only accepted the validity of the anthropic principle but also used it as an explanation for various aspects of the universe.

Modern anthropic arguments in physics and cosmology began in 1961, with a short paper in the British journal *Nature* by Robert Dicke. To understand Dicke's argument, we must go back to 1938, when the Nobel-Prize-winning physicist Paul Dirac pointed out that a certain combination of fundamental constants of nature, when multiplied and divided the right way, happened to equal the current age of the universe, about 10 billion years. (The fundamental constants of nature are such things as the speed of light, 186,000 miles per second, and the mass of an electron, 9.108×10^{-28} grams; such numbers are usually assumed to be the same everywhere and at all times. Indeed, every ray of light ever clocked in empty space travels at the same speed; every electron ever weighed has the same mass.) For Dirac, the coincidence between the two kinds of numbers—one built on microscopic quantities and the other on the universe as a whole—seemed too unlikely to be chance, and he suggested that there must be some link between the fundamental constants and the evolution of the universe. Since the age of the universe clearly increases in time, the fundamental constants of nature would also have to change in time to maintain the Dirac relationship.

Dicke explained Dirac's coincidence in a completely different way. Physicists, he reasoned, can exist only during a narrow window of time in the evolution of the universe. The carbon in physicists' bodies required a star to forge it, so a universe inhabited by physicists and other living beings must be old enough to have made stars. On the other hand, if the universe were too old, the stars would have burned out, eliminating the central source of heat and light that makes their orbiting planets habitable. Putting these limits together, physicists can exist only during an epoch when the age of the universe is approximately the lifetime of an average star. Dicke calculated this last quantity in terms of basic principles of physics and found it equal to the same combination of fundamental constants of nature that Dirac had noticed, numerically equal to about 10 billion years. Thus, the equality of Dirac's two numbers was not an accidental coincidence but a necessity for our existence. Dicke declared the constants to be constant, as usually assumed. Much earlier or much later than the present time, Dirac's combination of fundamental constants would not equal the age of the universe, but physicists wouldn't be here to discuss the situation.

Dicke's argument exemplified the weak anthropic principle. As an interesting footnote, Dirac published a short reply to Dicke's paper, saying that Dicke's analysis was sound but that he, Dirac, "preferred" his own argument because it allowed for the possibility that planets "could exist indefinitely in the future and life need never end." Aside from Dirac's objection, Dicke's weak anthropic principle is accepted by most cosmologists. Its statement and application concern only our actual universe.

In 1968 Brandon Carter, then at Cambridge University, stated the more controversial strong anthropic principle: The values of many of the fundamental constants of nature must lie within a restricted range in order to allow life to emerge at all, even during Dicke's window of time. Carter argues, for example, that the emergence of life requires the formation of planets, which, in turn, may require the existence of stars that can shed spinning pieces of themselves. When Carter analyzes the conditions necessary to form such "convective" stars, he finds that the values of some of the fundamental constants are restricted to a certain range. Such a range, of course, includes the values of *our* universe. Implicit in such an argument, and in all statements of the anthropic principle, is the assumption that galaxies and stars and other special conditions are necessary for life. Such an assumption is difficult to test, since our experience with life is limited to terrestrial biology.

In recent years, leading cosmologists have used the strong anthropic principle to explain certain properties of the universe. According to this line of thinking, only a universe with the specific properties ours has—including the values of certain fundamental constants and initial conditions—could allow our existence. For example, the ratio of the mass of the proton to the electron, observed in the lab to be about 2,000, could not be 2 or 2,000,000 in *our* universe because such values would produce a physics and chemistry and biology incompatible with living substance. Values of 2 or 2 million might exist in other universes, perfectly satisfactory in every way except in their ability to allow our existence. Out of all these other possible universes, only a small fraction would have a proton-to-electron mass ratio suitable for life. As another example, the strong anthropic principle was used by C. B. Collins and Hawking in 1973 to discuss the flatness problem. Starting with the assumption that galaxies and stars are necessary for life, they argued that a universe beginning with too much gravitational energy would collapse before it could form stars, and a universe beginning with too little would never allow the gravitational condensation of galaxies and stars. Thus, out of many possible universes, with many different initial values of omega, only in a universe where the initial value of omega was almost precisely 1 could we have existed. This anthropic solution to the flatness problem is accepted by some scientists and rejected by others.

Underlying these strong anthropic arguments is the notion that some properties of our universe are not fundamental and thus do not require fundamental explanations in terms of the laws of physics; instead, these properties are simply accidental outcomes drawn from a large hat. The numbers of our particular universe are what they are because anything much different would exclude us, and our existence is a given. Other numbers are characteristic of other universes, but we could not exist in most of them. The logic is similar to that used to explain why it rains on the days the parent drives his child to school and not on the days the child walks.

Scientists argue with each other about whether the anthropic principle constitutes a valid *explanation* of nature. For one thing, the explanations have force only if we invoke a large range of possible universes, with widely varying properties. Otherwise, we are back to explaining why the one and only nature is what it is. However, as mentioned earlier, some scientists are uncomfortable with postulating different universes. We inhabit just our one universe, and arguments that must go outside of that universe in order to explain it may have also gone outside science. For another thing, most scientists would rather explain nature in terms of basic laws that they can derive and prove, instead of conjectures about possibilities that might be right but can never be proved. To many scientists, a fundamental theory showing that, regardless of whether or not life exists, the ratio of the mass of the proton to the electron *must* be 1836 and nothing else—just as the ratio of a circle's circumference to its radius must be 3.1416 in Euclid's geometry—would be much preferred to an anthropic explanation. Some scientists consider that anthropic explanations of cosmic puzzles are proposed only when no better explanations can be found. Finally, the anthropic principle is sometimes tinged with teleological and religious suggestions. If life is indeed rare in all possible universes, was our particular universe designed with a purpose? Such questions, of course, run counter to the notion that our universe was simply an accident.

On the surface at least, anthropic arguments seem to involve life in a primary way. But there is a related question that sidesteps the issue of life. Could a universe exist with very different laws of physics, very different values for the speed of light and the mass of the electron and the initial value of omega? Or is ours the only universe possible, the only self-consistent set of laws and parameters? For some physicists, the search for the laws in our one universe has become a search after this much larger question.

CHAPTER • TWELVE

At Home

It is a cold, clear day in December. Snow fell through the night. This morning my daughter, age ten, and I are up early, in her room, looking outside at the white and talking softly. She has on flannel pajamas, holds our Maine coon cat, and leans against the window sill. I hold a cup of coffee. Do you think there are living creatures on other planets? she asks. Probably, I answer.

The air outside the window shines with two lights, one from the sun above and one from the sun below, reflected off the snow. And higher up, the air turns slowly into brilliant blue, which goes up and up as far as the eye can see. We are having a seeing contest.

Space is weird, says my daughter. You just can't imagine something that goes on forever. And if it doesn't go on forever, what's outside of it? I nod my head in agreement, and we tiptoe downstairs, to begin breakfast and make ready for the day.

Notes

Glossary

Acknowledgments

Illustration Credits

Index

Notes

TWO • EARLY COSMOLOGIES

Enuma Elish in A. Heidel, trans., *The Babylonian Genesis* (Chicago: University of Chicago Press, 1951). See also Thorkild Jacobsen, "Enuma Elish—The Babylonian Genesis," in Milton K. Munitz, ed., *Theories of the Universe* (New York: The Free Press, 1957).

Translation of Anaximander's writings in Thomas L. Heath, *Greek Astronomy,* Library of Greek Thought (London: Dent, 1932), pp. 5–7. See also the article on Anaximander by L. Taran in Charles C. Gillispie, ed., *Dictionary of Scientific Biography* (New York: Charles Scribner's Sons, 1981).

Article on Democritus by G. B. Kerferd in Gillispie, *Dictionary.*

Lucretius quotations from *De Rerum Natura,* trans. W. H. D. Rouse and Martin F. Smith (Cambridge: Harvard University Press, 1982), pp. 15 and 21.

Aristotle quotation from *On the Heavens,* trans. W. K. C. Guthrie (Cambridge: Harvard University Press, 1971), p. 23.

Kepler quotation from *Kepler's Conversations with Galileo's Sideral Messenger* (1610), trans. Edward Rosen (New York: Johnson Reprint Corporation, 1965), p. 43.

Copernicus quotation from *On the Revolutions of Heavenly Spheres* (1543), trans. Charles Glenn Wallis in *Great Books of the Western World* (Chicago: Encyclopedia Britannica, 1987), vol. 16, p. 520.

Portions of Digges' writings and commentary by Francis R. Johnson in Munitz, *Theories,* pp. 184–189.

Newton quotation from *Principia,* General Scholium, trans. Andrew Motte and Florian Cajori (Berkeley: University of California Press), vol. 2, pp. 544.

Newton's letters to Bentley in Munitz, *Theories.*

Einstein's pioneering paper in cosmology is Albert Einstein, "Cosmological Considerations on the General Theory of Relativity," *Sitzungsberichte der Preussischen Akad. d. Wissenschaften* 142 (1917). English trans., *The Principle of Relativity,* trans. W. Perrett and G. B. Jeffrey (New York: Dover, 1952). Einstein's new theory of gravity was first published in Albert Einstein, "The General Theory of Relativity," *Sitzungsberichte der Preussischen Akad. d. Wissenschaften* 778 (1915).

The astronomer Wilhelm de Sitter's concern about theorizing on the long-term evolution of the cosmos based upon our observational "snapshot" is evident, for example, in the first paragraphs of Wilhelm de Sitter, "On Einstein's Theory of Gravitation and Its Astronomical Consequences," *Monthly Notices of the Royal Astronomical Society* 78 (1917): 3.

Friedmann's nonstatic cosmological model is Alexander Friedmann, "On the Curvature of the World," *Zeitschrift fur Physik* 10 (1922): 377. Lemaître's nonstatic cosmological model is found in Georges Lemaître, "A Homogeneous Universe of Constant Mass and Increasing Radius Accounting for the Radial Velocity of Extra-Galactic Nebulae," *Annals of the Scientific Society of Brussels* 47A (1927): 49. English trans. in *Monthly Notices of the Royal Astronomical Society* 91 (1931): 483.

Einstein's published responses to Friedmann's nonstatic cosmological model are Albert Einstein, "Remark on the Paper by A. Friedmann, 'On the Curvature of the World,' " *Zeitschrift fur Physik* 11 (1922): 326; "Note on the Paper by A. Friedmann, 'On the Curvature of the World,' " *Zeitschrift fur Physik* 16 (1923): 228. In a hand-written draft of this second paper is a crossed-out sentence fragment saying that, to Friedmann's time-dependent solution of the cosmological equations, "a physical significance can hardly be ascribed." (The Collected Papers of Albert Einstein, unpublished document 1-026; quoted with permission of the Hebrew University of Jerusalem.)

Leavitt's work on Cepheid variables is in Henrietta S. Leavitt, Harvard Circular No. 173 (1912); reprinted in Harlow Shapley, ed., *Source Book in Astronomy* (Cambridge: Harvard University Press, 1966).

The first good measurement of the size of the Milky Way, using Cepheid variables, is reported in Harlow Shapley, "Studies Based on the Colors and Magnitudes in Stellar Clusters VI. On the Determination of the Distances of Globular Clusters," *Astrophysical Journal* 48 (1918): 89. Hubble's measurement of the distance to the Andromeda nebula is reported in Edwin P. Hubble, Annual Report of the Mount Wilson Observatory (1924), and published in "N.G.C. 6822, a Remote Stellar System," *Astrophysical Journal* 62 (1925): 409.

Hubble's discovery of the expansion of the universe is reported in Edwin P. Hubble, "A Relation between Distance and Radial Velocity among Extra-Galactic Nebulae," *Proceedings of the National Academy of Science* 15 (1929): 168.

The first measurements of the recessional motions of the nebulae were made by Vesto M. Slipher and reported in Arthur S. Eddington, *The Mathematical Theory of Relativity* (Cambridge: Cambridge University Press, 1924), p. 162.

The age of the universe cannot be exactly determined by the current rate of expansion of the universe because that rate has not been constant in time. However, the current rate gives a good estimate. To determine the age exactly, both the expansion rate and the change in the expansion rate must be known. The latter is determined by a quantity called omega, to be discussed.

FOUR • THE BIG BANG MODEL

Various kinds of evidence pointing to an open universe can be found in, for example, J. Richard Gott, III, James E. Gunn, David N. Schramm, and Beatrice M. Tinsley, "An Unbound Universe?" *Astrophysical Journal* 194 (1974): 543.

Empirical relationships between the intrinsic luminosity of a galaxy and the orbital speeds of its stars were discovered by Sandra M. Faber and Robert E. Jackson, "Velocity Dispersions and Mass-to-Light Ratios for Elliptical Galaxies," *Astrophysical Journal* 204 (1976): 668; R. B. Tully and J. P. Fisher, "A New Method of Determining Distances to Galaxies," *Astronomy and Astrophysics* 54 (1977): 661.

Hubble's program to measure the cosmic expansion rate at greater and greater distances is reported in Edwin P. Hubble and M. L. Humason, "The Velocity–Distance Relation among Extra-Galactic Nebulae," *Astrophysical Journal* 74 (1931): 43. Sandage's continuation of that program is reported in M. L. Humason, N. U. Mayall, and Allan R. Sandage, "Redshifts and Magnitudes of Extra-Galactic Nebulae," *Astronomical Journal* 61 (1956): 97.

If the universe were not expanding, then the size of the observable universe, or the horizon, would be exactly the distance light has traveled since the big bang. However, the universe is expanding. Thus, the *current* distance of an object is somewhat larger than the distance light travels to get from there to here because the object has moved farther away during the voyage of the light ray. This effect is not large. Taking it into account, the horizon is between 1 and 2 times the distance light has traveled since the big bang. If the universe is 10 billion years old, for example, the furthest point that can now be seen is between 10 and 20 billion light years away.

All quantitative estimates in this book assume that the universe is flat and has an expansion rate of 1 per 10 billion years. The estimates would not be substantially changed by other assumptions within the bounds of experimental evidence.

Theoretical calculations of the expected fraction of helium produced in nuclear reactions in the first few minutes after the big bang were done by Fred Hoyle and Roger J. Tayler, "The Mystery of the Cosmic Helium Abundance," *Nature* 203 (1964): 1108; Yakov B. Zel'dovich, "The Theory of the Expanding Universe as Originated by A. Friedmann," *Usp. Fig. Nauk* 80 (1963): 357, English trans. *Soviet Physics-Uspekhi* 6 (1964): 475; Yakov B. Zel'dovich, "Survey of Modern Cosmology," *Advances in Astronomy and Astrophysics* (New York: Academic Press, 1965), vol. 3; P. James E. Peebles, "Primordial He Abundance and Fireball II," *Astrophysical Journal* 146 (1966): 542. The expected abundances of helium and other light elements produced in the big bang were calculated by Robert V. Wagoner, William A. Fowler, and Fred Hoyle, "On the Synthesis of Elements at Very High Temperatures," *Astrophysical Journal* 148 (1967): 3. Refined calculations of the amount of lithium produced in the big bang were done by J. Yang, Michael S. Turner, Gary Steigman, David N. Schramm, and Keith Olive, "Primordial Nucleosynthesis: A Critical Comparison of Theory and Observation," *Astrophysical Journal* 281 (1984): 493. Also in support of the big bang model is the agreement between theory and observation for the abundance of deuterium, a form of hydrogen with one proton and one neutron. See Richard I. Epstein, J. M. Lattimer, and David N. Schramm, "The Origin of Deuterium," *Nature* 263 (1974): 198.

The cosmic background radiation was predicted by Ralph Alpher and Robert C. Herman, "Evolution of the Universe," *Nature* 162 (1948): 774; "On the Relative Abundance of the Elements," *Physical Review D* 74 (1948): 1737; Ralph Alpher, Robert C. Herman, and George A. Gamow, "Thermonuclear Reactions in the Expanding Universe," *Physical Review D* 74 (1948): 1198; George Gamow, "The Evolution of the Universe," *Nature* 162 (1948): 680; Robert H. Dicke, P. James E. Peebles, P. G. Roll, and David T. Wilkinson, "Cosmic Blackbody Radiation," *Astrophysical Journal* 142 (1965): 414.

The cosmic background radiation was discovered by Arno A. Penzias and Robert W. Wilson, "A Measurement of Excess Antenna Temperature at 4080 Mc/s," *Astrophysical Journal* 142 (1965): 419.

Milne's discussion of the cosmological principle is in Edward A. Milne, "World-Structure and the Expansion of the Universe," *Zeitschrift fur Astrophysik* 6 (1933): 1; "A Newtonian Expanding Universe," *Quarterly Journal of Mathematics (Oxford)* 5 (1934): 64.

A good discussion of the question of why scientists didn't immediately follow up on the original (1948) predictions of the cosmic background radiation can be found in Steven Weinberg, *The First Three Minutes* (New York: Basic Books, 1977), chap. 6.

The quote from Alpher and Hermann on the steady state model comes from Ralph A. Alpher and Robert C. Herman, "Reflections of Early Work on Big Bang Cosmology," *Physics Today* 41, no. 8 (1988): 26.

FIVE • OTHER COSMOLOGICAL MODELS

The quote from Tolman on the application of thermodynamics to cosmology comes from Richard C. Tolman, *Relativity, Thermodynamics, and Cosmology* (Oxford: Clarendon Press, 1934), p. 444.

The quote from Dicke and collaborators on the oscillating universe comes from Dicke, Peebles, Roll, and Wilkinson, "Cosmic Blackbody Radiation," p. 415.

The steady state model was proposed by Herman Bondi and Thomas Gold in "The Steady-State Theory of the Expanding Universe," *Monthly Notices of the Royal Astronomical Society* 108 (1948): 252; Fred Hoyle, "A New Model for the Expanding Universe," *Monthly Notices of the Royal Astronomical Society* 108 (1948): 372. The quote from Bondi and Gold on the steady state model comes from their 1948 paper, p. 254.

Schmidt's discovery of quasars was reported in Maarten Schmidt, "3C 273: A Star-like Object with Large Redshift," *Nature* 197 (1963): 1040.

Rees and Sciama's analysis of quasar data is in Dennis W. Sciama and Martin J. Rees, "Cosmological Significance of the Relation between Redshift and Flux Density for Quasars," *Nature* 211 (1965): 1283.

SIX • DIFFICULTIES WITH THE BIG BANG MODEL

Misner's statement of the horizon problem is in Charles W. Misner, "The Mixmaster Universe," *Physical Review Letters* 22 (1969): 1071.

The analogy using a rock thrown upward from the earth neglects the effects of air resistance on the rock.

Dicke's statement of the flatness problems is in Robert H. Dicke, *Gravitation and the Universe,* The Jayne Lectures for 1969 (American Philosophical Society, 1969), p. 62. A few years later, the British cosmologists discussed the flatness problem in the context of the anthropic principle. See C. B. Collins and Stephen W. Hawking, "Why Is the Universe Isotropic?" *Astrophysical Journal* 180 (1973): 317; Brandon Carter, "Large Number Coincidences and the Anthropic Principle in Cosmology," in M. S. Longair, ed., *Confrontation of Cosmological Theories with Observational Data,* IAU Symposium 63 (Dordrecht: Reidel, 1974). The influential restatement of the flatness problem occurs in Robert H. Dicke and P. James E. Peebles, "The Big Bang Cosmology—Enigmas and Nostrums," in Stephen W. Hawking and Werner Israel, eds., *General Relativity: An Einstein Centenary Survey* (Cambridge: Cambridge University Press, 1979).

Guth's paper on the inflationary universe model is Alan Guth, "Inflationary Universe: A Possible Solution to the Horizon and Flatness Problems," *Physical Review D* 23 (1981): 347.

Penrose's papers mentioning the entropy problem in cosmology include Roger Penrose, "Singularities in Cosmology," in Longair, *Confrontation;* "Singularities and Time-Asymmetry," in Hawking and Israel, *Einstein Centenary.*

SEVEN • LARGE-SCALE STRUCTURE AND DARK MATTER

Charlier's papers on a hierarchical cosmos include C. V. L. Charlier, "The Planetary Rotation Problem," *Ark. Mathematics, Astronomy, and Physics* 4 (1908): 1; "System of an Infinite Universe," *Ark. Mathematics, Astronomy, and Physics* 16 (1922): 1. Shapley's suggestion of an evolutionary tendency in the groupings of galaxies is in Harlow Shapley, "Note on the Distribution of Remote Galaxies and Faint Stars," *Harvard Bulletin* 890 (1933): 1. Zwicky's suggestion about clusters of galaxies is in Fritz Zwicky, "On the Clustering of Nebulae," *Publications of the Astronomical Society of the Pacific* 50 (1938): 218. De Vaucouleurs's discovery of a supercluster of galaxies is in Gerard de Vaucouleurs, "Evidence for a Local Supergalaxy," *Astronomical Journal* 58 (1953): 30. De Vaucouleurs's proposal of an inhomogeneous and hierarchical universe is in Gerard de Vaucouleurs, "The Case for Hierarchical Clustering," *Science* 167 (1970): 1203.

Observational work during the period 1975 to 1990 on the distribution of galaxies includes G. Chincarini and H. J. Rood, "Empirical Properties of the Mass Discrepancy in Groups and Clusters of Galaxies. IV. Double Compact Galaxies," *Nature* 257 (1975): 294; W. G. Tifft and S. A. Gregory, "Observations of the Large-Scale Distribution of Galaxies," in M. S. Longair and J.

Einasto, eds., *The Large Scale Structure of the Universe,* IAU Symposium 79 (Dordrecht: Reidel, 1978); M. Joeveer and J. Einasto, "Has the Universe a Cell Structure?" in Longair and Einasto, *Large Scale Structure;* S. A. Gregory and L. A. Thompson, "The Coma/A 1367 Supercluster and Its Environs," *Astrophysical Journal* 222 (1978): 784; Robert P. Kirshner, Augustus Oemler, Jr., Paul L. Schechter, and Stephen A. Schectman, "A Million Cubic Megaparsec Void in Bootes?" *Astrophysical Journal Letters* 248 (1981): L57; S. A. Gregory, L. A. Thompson, and W. G. Tifft, "The Perseus Supercluster," *Astrophysical Journal* 243 (1981): 411; H. P. Haynes and R. Giovanelli, "A 21 Centimeter Survey of the Perseus–Pisces Supercluster. I. The Declination Zone +27.5 to 33.5 Degrees," *Astronomical Journal* 90 (1985): 2445; Valerie de Lapparent, Margaret J. Geller, and John P. Huchra, "A Slice of the Universe," *Astrophysical Journal Letters* 302 (1986): L1; Margaret J. Geller and John P. Huchra, "Mapping the Universe," *Science* 246 (1989): 897; T. J. Broadhurst, R. S. Ellis, D. C. Koo, and A. S. Szalay, "Large-Scale Distribution of Galaxies at the Galactic Poles," *Nature* 343 (1990): 726.

Peebles's discussion of the homogeneity of the universe can be found in P. James E. Peebles, *The Large Scale Structure of the Universe* (Princeton: Princeton University Press, 1980), p. 10. Here, Peebles refers to work on the radio and X-ray background of A. Webster, "The Clustering of Radio Sources II. The 4C, GB, and MC1 Surveys," *Monthly Notices of the Royal Astronomical Society* 175 (1976): 71; P. James E. Peebles, "Stability of a Hierarchical Clustering Pattern in the Distribution of Galaxies," *Astronomy and Astrophysics* 68 (1978): 345; A. M. Wolfe, "New Limits on the Shear and Rotation of the Universe from the X-ray Background," *Astrophysical Journal Letters* 159 (1970): L61; Andrew C. Fabian, "Analysis of X-ray Background Fluctuations," *Nature* 237 (1972): 19.

Vera Rubin's earlier work on peculiar velocities is in Vera C. Rubin, "Differential Rotation of the Inner Metagalaxy," *Astronomical Journal* 56 (1951): 47; Vera C. Rubin, W. K. Ford, Jr., and J. S. Rubin, "A Curious Distribution of Radial Velocities of ScI Galaxies with $14.0 \leq m \leq 15.0$," *Astrophysical Journal Letters* 183 (1973): L111; Vera C. Rubin, N. Thonnard, W. K. Ford, Jr., and M. S. Roberts, "Motion of the Galaxy and the Local Group Determined from the Velocity Anisotropy of Distant Sc I Galaxies. II. The Analysis for the Motion," *Astronomical Journal* 81 (1976): 719

Peculiar velocities of galaxies toward the Great Attractor were first reported by Alan Dressler, Sandra M. Faber, D. Burstein, R. L. Davies, Donald Lynden-Bell, R. J. Terlevich, and Gary Wegner (the Seven Samurai), "Spectroscopy and Photometry of Elliptical Galaxies: A Large-Scale Streaming Motion in the Local Universe," *Astrophysical Journal Letters* 313 (1987): L37.

The recent theoretical work that computes the gravitational field and distribution of mass from the observed peculiar velocities of galaxies is Edmund Bertschinger and Avishai Dekel, "Recovering the Full Velocity and Distance Fields from Large-Scale Redshift-Distance Samples," *Astrophysical Journal Letters* 336 (1989): L5; Avishai Dekel, Edmund Bertschinger, and Sandra M. Faber, "Potential, Velocity, and Density Fields from Sparse and Noisy Red-Shift Distance Samples: Method," *Astrophysical Journal* 364 (1990): 349; Edmund Bertschinger, Avishai Dekel, Sandra M. Faber, Alan Dressler, and David Burstein, "Potential, Velocity, and Density Fields from Red-Shift Distance Samples: Application, Cosmography Within 6000 Kilometers per Second," *Astrophysical Journal* 364 (1990): 370.

Dark matter was first discovered by Fritz Zwicky, "Spectral Displacement of Extra-Galactic Nebulae," *Helvetian Physics Acta* 6 (1933): 110. Ostriker and Peebles's theoretical work on dark matter in rotating galaxies is in Jeremiah P. Ostriker and P. James E. Peebles, "A Numerical Study of the Stability of Flattened Galaxies: Or, Can Cold Galaxies Survive?" *Astrophysical Journal* 186 (1973): 467. Theoretical calculations of the dark matter in a variety of astronomical systems from individual galaxies to large clusters of galaxies is in Jeremiah P. Ostriker, P. James E. Peebles, and Amos Yahil, "The Size and Mass of Galaxies and the Mass of the Universe," *Astrophysical Journal Letters* 193 (1974): L1; J. Einasto, A. Kaasik, and E. Saar, "Dynamic Evidence on Massive Coronas in Galaxies," *Nature* 250 (1974): 309. The influential early observations of dark matter are reported in Vera C. Rubin, W. K. Ford, Jr., and N. Thonnard, "Extended Rotation Curves of High-Luminosity Spiral Galaxies. IV. Systematic Dynamical Properties," *Astrophysical Journal Letters* 225 (1978): L107; Albert Bosma, "The Distribution and Kinematics of Neutral Hydrogen in Spiral Galaxies of Various Morphological Types," Ph.D. thesis, University of Groningen (1978).

"The first discovery of a gravitational lens was D. Walsh, R. F. Carswell, and R. J. Weymann, "0957 + 581 A, B: Twin Quasistellar Objects, or Gravitational Lens?" *Nature* 279 (1979): 381. Tyson's recent work in using gravitational lenses to map dark matter is in J. A. Tyson, F. Valdes, and R. A. Wenk, "Detection of Systematic Gravitational Lens Galaxy Image Alignments," *Astrophysical Journal Letters* 349 (1990): L1.

Jeans's calculation of the formation of mass concentrations is J. Jeans, "The Stability of a Spherical Nebula," *Philosophical Transactions of the Royal Society* 199A (1902): 49. Lemaître's sketch of the gravitational hierarchy model is in Georges Lemaître, "Spherical Condensations in the Expanding Universe," *Comptes Rendus de l'Academie des Sciences, Paris* 196 (1933): 903; Peebles's early work on the gravitational hierarchy model is in P. James E. Peebles, "The

Black-body Radiation Content of the Universe and the Formation of Galaxies," *Astrophysical Journal* 142 (1965): 1317.

Computer simulations using gas dynamics can be found in R. Y. Cen, A. Jameson, F. Liu, and J. P. Ostriker, "The Universe in a Box: Thermal Effects in the Standard Cold Dark Matter Scenario," *Astrophysical Journal Letters* 362 (1990): L41.

Silk's work on the effects of radiation on forming galaxies is in Joseph I. Silk, "Fluctuations in the Primordial Fireball," *Nature* 215 (1967): 115.

The pancake model was developed by Yakov B. Zel'dovich, "Gravitational Instability: An Approximate Theory for Large Density Perturbations," *Astronomy and Astrophysics* 5 (1970): 84; A. G. Doroshkevich, V. S. Ryabenki, and S. F. Shandarin, "Nonlinear Theory of Development of Potential Perturbations," *Astrofizika* 9 (1973): 257; A. G. Doroshkevich, Rashid A. Sunyaev, and Yakov B. Zel'dovich, in Longair, *Confrontation*.

Recent computer calculations of the cold matter model are in Changboom Park, "Large N-Body Simulations of a Universe Dominated by Cold Dark Matter," *Monthly Notices of the Royal Astronomical Society* 242 (1990): 59p; also recent calculations not yet published by Edmund Bertschinger and James Gelb at MIT, Richard Gott and Changboom Park at Princeton, and Jens Villumsen at Ohio State University.

The new, 2.5 million galaxy survey is reported in S. J. Maddox, G. Efstathiou, W. Sutherland, and J. Loveday, *Monthly Notices of the Royal Astronomical Society* 242 (1990): 43. The new galaxy survey by scientists from Queen Mary and Westfield College, University of Durham, University of Oxford, and University of Toronto is W. Saunders, C. Frenk, M. Rowan-Robinson, G. Efstathiou, A. Lawrence, N. Kaiser, R. Ellis, J. Crawford, X. Xia, and I. Parry, "The Density Field of the Local Universe," *Nature* 349 (1991): 32. The work by Bahcall and Soneira is reported in N. A. Bahcall and R. M. Soneira, "The Spatial Correlation Function of Rich Clusters of Galaxies." *Astrophysical Journal* 270 (1983): 20.

Recent measurements of the variations in the cosmic background radiation, establishing an upper limit of several parts in 100,000, have been done by C. R. Lawrence, A. Readhead, and S. T. Myers at the California Institute of Technology; P. M. Lubin and P. Meinhold at the University of California at San Diego; and S. P. Boughn, E. S. Cheng, D. A. Cottingham, and D. J. Fixsen from Haverford, the Goddard Space Flight Center, and Berkeley.

Models of explosive galaxy formation can be found in A. G. Doroshkevich, Yakov B. Zel'dovich, and I. D. Novikov, "The Origin of Galaxies in an

Expanding Universe," *Soviet Astronomy AJ* 11 (1967): 233; Jeremiah P. Ostriker and Lenox L. Cowie, "Galaxy Formation in an Integalactic Medium Dominated by Explosions," *Astrophysical Journal Letters* 243 (1981): L127.

NINE • INITIAL CONDITIONS AND QUANTUM COSMOLOGY

Hawking and Hartle's calculations of the initial conditions of the universe are in Stephen W. Hawking, "The Boundary Conditions of the Universe," *Pontificae Academiae Scientarium Scripta Varia* 48 (1982): 563; James B. Hartle and Stephen W. Hawking, "Wave Function for the Universe," *Physical Review D* 28 (1983): 2960; Stephen W. Hawking, "The Quantum State of the Universe," *Nuclear Physics B* 239 (1984): 257.

Penrose and Hawking's work on the "singularity theorems" and the reality of an ultra-high-density beginning of the universe is in Roger Penrose, "Gravitational Collapse and Space-Time Singularities," *Physical Review Letters* 14 (1965): 57; Stephen W. Hawking and Roger Penrose, "The Singularities of Gravitational Collapse and Cosmology," *Proceedings of the Royal Society of London* A314 (1969): 529.

TEN • PARTICLE PHYSICS, THE NEW COSMOLOGY, AND THE INFLATIONARY UNIVERSE MODEL

Theoretical work on the relationship between the cosmic helium abundance and the number of lepton types can be found in Hoyle and Tayler, *Mystery;* Robert V. Wagoner, "Cosmological Element Production," *Science* 155 (1967): 1369; V. F. Shvartsman, "Density of Relic Particles with Zero Rest Mass in the Universe," *Soviet Physics JETP Letters* 9 (1969): 184; Gary Steigman, David N. Schramm, and James E. Gunn, "Cosmological Limits to the Number of Massive Leptons," *Physics Letters* 66B (1977): 202.

The experiments at CERN to reveal the number of types of leptons are reported in P. Aarnio et al., "Measurement of the Mass and Width of the Z° Particle from Multihadronic Final States Produced in $e^{+}e^{-}$ Annihilations," *Physics Letters B* 231 (1989): 539; B. Adeva et al., "A Determination of the Properties of the Neutral Intermediate Vector Boson Z°," *Physics Letters B* 231 (1989): 509; M. Z. Akrawy, et al., "Measurement of the Z° Mass and Width with the Opal Detector at LEP," *Physics Letters B* 231 (1989): 530; D. Decamp et al., "Determination of the Number of Light Neutrino Species," *Physics Letters B* 231 (1989): 519. The experiments at the Stanford Linear Accelerator are reported in G. S. Abrams et al., "Searches for New Quarks and Leptons Produced in Z-Boson Decay," *Physical Review Letters* 63 (1989): 2447.

One of the first and simplest grand unified theories, called SU(5), was that of Howard Georgi and Sheldon L. Glashow, "Unity of All Elementary Particle Forces," *Physical Review Letters* 32 (1974): 438.

A good review of cosmic strings is Alexander Vilenkin, "Cosmic Strings," *Scientific American* 257 (1987): 94; for a more technical review, see A. Vilenkin, "Cosmic Strings and Domain Walls," *Physics Reports* 121 (1985): 263.

Theoretical calculations of the ratio of photons to baryons include S. Dimopoulos and L. Susskind, "Baryon Number of the Universe," *Physical Review D* 18 (1978): 4500; A. D. Dolgov, "Baryon Asymmetry of the Universe and Thermodynamical Equilibrium Disturbance," *Zh. Eksp. Teor. Pis'ma Red.* 29 (1979): 254; J. Ellis, M. K. Gaillard, and D. V. Nanopoulos, "Baryon Number Generation in Grand Unified Theories," *Physics Letters B* 80 (1979): 360; A. Yu Ignatiev, N. V. Krashikov, V. A. Kuzmin, and A. N. Tavkhelidze, "Universal CP-Noncovariant Superweak Interaction and Baryon Asymmetry of the Universe," *Physics Letters B* 76 (1978): 436; D. Toussaint, S. B. Trieman, Frank Wilczek, and A. Zee, "Matter-Antimatter Accounting, Thermodynamics, and Black-Hole Radiation," *Physical Review D* 19 (1979): 1036; Steven Weinberg, "Cosmological Production of Baryons," *Physical Review Letters* 42 (1979): 850; M. Yoshimura, "Unified Gauge Theories and the Baryon Number of the Universe," *Physical Review Letters* 41 (1978): 281.

Early papers containing ingredients of the inflationary universe model include R. Brout, F. Englert, and P. Spindel, "Cosmological Origin of the Grand-Unification Mass Scale," *Physical Review Letters* 43 (1979): 417; Demos Kazanas, "Dynamics of the Universe and Spontaneous Symmetry Breaking," *Astrophysical Journal Letters* 241 (1980): L95; Martin B. Einhorn and K. Sato, "Monopole Production in the Very Early Universe in a First-Order Phase Transition," *Nuclear Physics B* 180 (1981): 385; A. Starobinsky, "Spectrum of Relict Gravitational Radiation and the Early State of the Universe," *Soviet Physics JETP Letters* 30 (1979): 682; "A New Type of Isotropic Cosmological Models without Singularity," *Physics Letters* 91B (1980): 99. Guth's paper on the inflationary universe model is Alan Guth, "Inflationary Universe: A Possible Solution to the Horizon and Flatness problems," *Physical Review D* 23 (1981): 347. Linde, Albrecht, and Steinhardt's new version of the inflationary universe model is in Andrei D. Linde, "A New Inflationary Universe Scenario: A Possible Solution of the Horizon, Flatness, Homogeneity, Isotropy, and Primordial Monopole Problems," *Physics Letters B* 108 (1982): 389; Andreas Albrecht and Paul J. Steinhardt, "Cosmology for Grand Unified Theories with Radiatively Induced Symmetry Breaking," *Physical Review Letters* 48 (1982): 1220. Linde's chaotic inflationary universe model is in Andrei D. Linde, "Chaotic Inflation," *Physics Letters B* 129 (1983): 177; "Particle Physics and Inflationary Cosmology," *Physics Today*, no. 9, 40 (1987): 61.

Theoretical calculations suggesting problems with the new inflationary universe model include Alan H. Guth and S-Y Pi, "Fluctuations in the New Inflationary Universe," *Physical Review Letters* 49 (1982): 1110; Stephen W. Hawking, "The Development of Singularities in a Single Bubble Inflationary Universe," *Physics Letters* 115B (1982): 295.

ELEVEN • THE ANTHROPIC PRINCIPLE

Dicke's paper on the anthropic principle is Robert H. Dicke, "Dirac's Cosmology and Mach's Principle," *Nature* 192 (1961): 440. Dirac's reply to Dicke is Paul A. M. Dirac, "Reply to Dicke," *Nature* 192 (1961): 441.

Dirac's original paper on the coincidence of numbers in physics is Paul A. M. Dirac, "New Basis for Cosmology," *Proceedings of the Royal Astronomical Society of London* A165 (1938): 199. Dirac's special combination of fundamental constants is h^2/cGm_p^3, where h is Planck's constant of quantum physics, c is the speed of light, G is Newton's constant of gravity, and m_p is the mass of the proton. This is numerically equal to about 4×10^{15} seconds, or 10^8 years, which is considered very close to 10^{10} years by cosmological standards.

Carter's papers on the strong anthropic principle are Brandon Carter, "Large Numbers in Astrophysics and Cosmology," unpublished preprint from the Institute of Theoretical Astronomy, Cambridge University (1968); "Large Number Coincidences and the Anthropic Principle in Cosmology," in Longair, *Confrontation*. Applications of the strong anthropic principle to explain properties of the universe include the original papers by Carter; Collins and Hawking, "Why Is the Universe Isotropic?"; Bernard J. Carr and Martin J. Rees, "The Anthropic Principle and the Structure of the Physical World," *Nature* 278 (1979): 605; Steven Weinberg, "Anthropic Bounds on the Cosmological Constant," *Physical Review Letters* 59 (1987): 2607. The paper by Hawking and Collins uses the strong anthropic principle to solve the flatness problem.

Glossary

abundances The relative amounts of chemical elements. For example, hydrogen makes up about 75 percent of the mass of the universe, so its "cosmic abundance" is 75 percent.

anisotropy The condition in which the universe appears different in different directions.

anthropic principle The weak form of the anthropic principle states that life can exist only during a brief period of the history of our universe. The strong form of the principle states that out of all possible values for the fundamental constants of nature and the initial conditions of the universe, only a small fraction could allow life to form at all, at any time. (See boundary conditions; fundamental constants of nature.)

arcsecond A unit of angle in astronomy. One arcsecond is 1/3600 of a degree.

baryons Subatomic particles that interact via the strong nuclear force. The proton and the neutron are examples of baryons. Until recently, it was believed that baryons were conserved, that is, the net number of baryons before and after any physical process was unchanged. Grand unified theories (GUTs) of physics, first proposed in the mid 1970s, suggest that baryons may not be conserved.

baryon-to-photon ratio (See photon-to-baryon ratio.)

big bang model An evolutionary model of cosmology in which the universe began about 10 billion years ago, in a state of extremely high density and temperature. According to this model, the universe has been expanding, thinning out, and cooling since its beginning. It is an observational fact that distant galaxies are all moving away from our own galaxy, as predicted by the big bang model. (See closed universe; Friedmann models; open universe.)

blackbody radiation A unique type of radiation whose spectrum and other properties are completely characterized in terms of a single quantity, temperature. Blackbody radiation is produced after a group of particles and photons have come into thermal equilibrium with one another, with every reaction between the particles balanced by the reverse reaction, so that the system as a whole has stopped changing. In this situation, all parts of the system, including the radiation, have come to the same temperature. Blackbody radiation would be produced, for example, inside an oven that is maintained at a constant temperature and in which the door has been left closed for a long time. (See photon; spectrum; thermal equilibrium.)

black hole A mass that is sufficiently compact that not even light can escape its intense gravity. Thus it appears black from the outside. If the sun were compressed to a sphere about four miles in diameter, it would become a black hole. It is believed that some massive stars, after exhausting their nuclear fuel, collapse under their own weight to form black holes.

boundary conditions Conditions needed to determine the evolution of a physical system, given the laws of nature. For example, the swing of a pendulum is determined both by the laws of mechanics and gravity and by the initial height at which the pendulum is released. This latter is called a boundary condition, or an initial condition.

bubbles (in the large-scale distribution of galaxies) The name for the structures formed by the observed distribution of galaxies in space. Some surveys of nearby galaxies show that galaxies are located on roughly spherical shells, called bubbles, of about a hundred million light years in diameter (about a thousand times the diameter of a single galaxy). Few galaxies reside in the interior of a "bubble."

CCDs An acronym for charge coupled devices, which are highly sensitive photoelectric devices that can electronically record the intensity and point of arrival of tiny amounts of light. CCDs are placed at the receiving end of telescopes, to "take pictures" of very faint astronomical objects; they have almost completely replaced photographic plates.

Cepheid A type of star that oscillates in brightness, growing dim, then bright, then dim again in a cyclical way, with the cycle time closely correlated with the star's luminosity. Thus, by measuring the light cycle time of a Cepheid star, one can calculate its intrinsic luminosity. A comparison of a star's luminosity to its *apparent* brightness then gives the distance of the star. Cepheid stars are among the few astronomical objects whose distances can be reliably determined. (See luminosity.)

chaotic inflation A variation of the inflationary universe model in which random quantum fluctuations are continually forming new universes. (See inflationary universe model; new inflation; old inflation; quantum fluctuations; quantum mechanics.)

closed universe A universe that has a finite size. Closed universes expand for a finite time, reach a maximum size, and then collapse. In closed universes, the inward pull of gravity dominates and eventually reverses the outward flying apart of matter; that is, gravitational energy dominates the kinetic energy of expansion. The value of omega is greater than 1 for a closed universe. If a universe begins closed, it remains closed; if it begins open, it remains open; if it begins flat, it remains flat. In the big bang model of the universe, the question of whether the universe is closed, open, or flat is determined by the initial conditions, just as the fate of a rocket launched from earth is determined by its initial upward velocity relative to the strength of earth's gravitational pull. If the initial rate of expansion of the universe was lower than a critical value, determined by the mass density, the universe will expand only for a certain period of time and then collapse, just as a rocket launched with a velocity below a critical value, dependent on the strength of earth's gravity, will reach a maximum height and then fall back to earth. This is the behavior of a closed universe. If the initial rate of expansion of the universe was larger than a critical value, the universe is open and will keep expanding forever. If the initial rate of expansion was precisely the critical value, the universe is flat and will expand forever, but with a rate of expansion that approaches zero. (See flat universe; omega; open universe.)

cluster In cosmology, a group of galaxies closer together than could be expected if galaxies were randomly scattered through space. The size of a typical cluster is about 15 million light years. Clusters of galaxies can further bunch together to form a "supercluster" of galaxies, the typical size of which is about 150 million light years.

clustering In cosmology, the observed tendency of galaxies to bunch together, rather than to distribute themselves uniformly and independently of one another. (See galaxy.)

cold dark matter (See dark matter.)

cold dark matter model A leading theoretical model for explaining the clustering of galaxies and other large distributions of cosmic mass. The cold dark matter model is based on the inflationary universe model, assumes that the universe is flat, and assumes that the "missing mass" is composed of slowly moving particles that easily cluster. (See inflationary universe model; missing mass.)

Coma cluster A massive cluster of galaxies nearest to our home cluster, which is known as the local group. The Coma cluster, about 300 million light years from us, contains about 1,000 galaxies in a region about 10 million light years across. (See clustering; galaxy.)

Copenhagen interpretation of quantum mechanics The view of quantum mechanics holding that a physical system exists in one and only one of its possible states after a measurement is made. Prior to the measurement, the system has no physical existence and is describable only in terms of the probability of each possible result of a measurement. (See many-worlds interpretation of quantum mechanics; quantum mechanics.)

cosmic background radiation Often called simply background radiation, or cosmic microwave radiation, a pervasive bath of radio waves coming from all directions of space. According to the big bang theory, this radiation was produced by the collisions of particles when the universe was much younger and hotter, and it uniformly filled up space. The collisions between the radiation and matter stopped when the universe was about 300,000 years old, and the cosmic background radiation has been traveling free through space ever since. The cosmic background radiation is now in the form of radio waves.

cosmological constant A contribution to gravity that results from the effective mass density, or energy density, in the vacuum. A positive cosmological constant acts as if it were negative gravity—it makes two masses repel each other instead of attract each other. Einstein's first cosmological model contained a cosmological constant, which appeared as an additional term in the equations of general relativity. (See false vacuum; vacuum.)

cosmological principle The statement that the universe is homogeneous and isotropic on the large scale, that is, it appears the same at all places and, from any one place, looks the same in all directions.

critical mass density The value of average cosmic mass density above which the universe is closed. The average mass density of the universe is obtained by measuring the mass in a very large volume of space, including many galaxies, and dividing by the size of the volume. The critical mass density is determined by the current rate of expansion of the universe. According to estimates of the current rate of expansion, the current critical mass density is about 10^{-29} grams per cubic centimeter. According to the best measurements, the average mass density of our universe appears to be about one tenth the critical mass density. (See closed universe; omega; open universe.)

curvature The departure of the geometry of the universe from Euclidean (flat) geometry. Qualitatively, the curvature is indicated by the curvature parameter, denoted by k. The values $k = 0, 1, -1$ refer to flat (uncurved) geometry, closed geometry, and open geometry, respectively. In a flat geometry, for example, the circumference of a circle is twice pi times its radius. In a closed geometry, the circumference is smaller than twice pi times the radius; in an open geometry, it is larger. (See closed universe; flat universe; open universe.)

dark matter Matter in the universe that we detect by its gravitational influences, yet do not see. Dark matter that has small random speed and is easily concentrated by gravity is called cold dark matter. Dark matter that has large random speed and is thus able to resist gravitational clumping is called hot dark matter. Recent models to explain the observed pattern of galaxy clustering can be characterized, in part, as to whether they invoke hot dark matter or cold dark matter. However, since we do not know what the dark matter is, we do not have any direct evidence of whether it is cold or hot.

deceleration parameter A parameter that measures the rate of slowing down of the expansion of the universe. Gravity causes the slowing down. The deceleration parameter equals omega (another cosmological parameter) when the universe is dominated by radiation, approximately the first 100,000 years after the big bang, and 1/2 omega when the universe is dominated by matter. Since the deceleration parameter is equivalent to omega (assuming a cosmological constant of zero, as is often done), it determines the ultimate fate and spatial geometry of the universe. The deceleration parameter is often denoted by the symbol q_0. (See omega.)

density fluctuations Random inhomogeneities in an otherwise smooth distribution of matter.

density parameter Another name for omega. (See omega.)

de Sitter model A particular solution to Einstein's cosmological equations, found by Wilhelm de Sitter in 1917, in which space expands at a rapid, exponential rate. This solution was very different from the solutions of Friedmann and of Lemaître, in which the universe expands at a much slower rate (a rate with the distance between any two points increasing as something between the square root of time and linearly with time). The Friedmann and Lemaître type solutions became incorporated in the standard big bang model. Recent modifications of the big bang model, such as the inflationary universe model, propose that the universe went through a period of exponential growth, or a de Sitter phase, early in its evolution.

deuterium An atomic nucleus consisting of a proton and a neutron. It is believed that deuterium was the first compound nucleus formed in the infant universe.

differential equation An equation that describes the evolution of a system over time, given boundary conditions for the system. Almost all of the laws of physics are expressed in the mathematics of differential equations. (See boundary conditions.)

Dirac large-number hypothesis The current age of the universe, divided by the time it takes light to cross the radius of a proton, is about 10^{40}. This number is also approximately equal to the ratio of the strengths of the electromagnetic and gravitational forces. Dirac felt that the approximate equality of these two large numbers was too unlikely to be accidental and that some physical process must be at work to maintain the equality. Since the first number clearly changes in time (because the age of the universe is increasing), Dirac proposed that the "fundamental constants of nature" entering the second number should also change in time, to maintain the equality.

Einstein-de Sitter theory A particular solution to Einstein's cosmological equations in which the universe is flat. (See flat universe.)

Einstein equations The equations of Einstein's theory of gravity, called general relativity. The Einstein equations quantitatively specify the gravity produced by matter and energy. Since gravity is believed to be the principal force acting over very large distances, the Einstein equations are used in modern theories of cosmology.

electromagnetic force One of the four fundamental forces of nature. Electricity and magnetism arise from the electromagnetic force. The other three fundamental forces are the gravitational force, the weak nuclear force, and the strong nuclear force. (See gravitational force; nuclear force.)

electron volt A unit of energy or of mass. The electron weighs about 10^{-27} grams, which is equivalent to about 500 million (5×10^{8}) electron volts of energy. Thus, the electron volt is tiny by ordinary standards. The energy released by dropping a penny (3 grams) to the floor is about 4×10^{17} electron volts.

electroweak theory The theory that unifies the electromagnetic force and the weak nuclear force into a single force. This theory was developed in the 1960s by Sheldon Glashow, Steven Weinberg, and Abdus Salam and has been subsequently confirmed in the laboratory.

ensemble (of universes) A hypothetical group of many universes of varying properties. Some physicists attempt to estimate how "probable" are the properties of our universe by imagining it as a sample from an ensemble of universes.

entropy A quantitative measure of the degree of disorder of a physical system. Highly disordered systems have high entropy; highly ordered systems have low entropy. One of the laws of physics, the second law of thermodynamics, says that the entropy of any isolated physical system can only increase in time.

equivalence principle The statement that a gravitational force is completely equivalent in all of its physical effects to an overall acceleration in the opposite direction. For example, a person in an elevator in space accelerating upward at 32 feet per second per second would feel the floor pushing upward against her feet in exactly the same way as if the elevator were at rest on earth, where gravity pulls downward with an acceleration of 32 feet per second per second. The "weak equivalence principle," which is not as strong as the equivalence principle, states that all objects, independent of their mass or composition, fall with the same acceleration in the presence of gravity. The Eötvös experiment, and later refinements of this experiment, have proven the weak equivalence principle.

Euclidean geometry The geometry developed by the Greek Euclid about 300 B.C. Euclidean geometry, like all geometries, deduces certain results from a set of starting assumptions. One of the critical assumptions of Euclidean geometry is that, given any straight line and a point not on that line, there is exactly one line that can be drawn through that point parallel to the first line. One of the results of Euclidean geometry is that the interior angles of any triangle add up to 180 degrees. Euclidean geometry is the geometry we learn in high school.

explosive galaxy formation A theory of galaxy formation wherein the explosion of a large number of stars creates a giant shock wave that travels outward and compresses the surrounding gas. Galaxies form in the regions of high-density gas.

exponential expansion Extremely rapid expansion. For example, a balloon that doubles its size every second, so that it is one inch after one second, two inches after two seconds, four inches after three seconds, and eight inches after four seconds, is expanding exponentially. By contrast, a balloon whose radius is one inch after one second, two inches after two seconds, three inches after three seconds, and four inches after four seconds is expanding linearly with time, rather than exponentially. According to the inflationary universe model, the early universe went through a brief period of exponential expansion, during which its size increased enormously.

extragalactic distance scale The set of distances to astronomical objects outside our galaxy. It is difficult to obtain distances to objects further than about 10 million light years with accuracies better than about 25 percent.

Faber-Jackson relation An empirically observed correlation between the speeds of stars in the center of a galaxy and the luminosity of the galaxy—the higher the random speeds, the more luminous the galaxy. Since the speeds of stars can be directly measured by the Doppler shift in their colors, the Faber-Jackson relation permits an estimate of the luminosity of a galaxy. By comparing this with the *observed* brightness, we can infer the distance to the galaxy.

false vacuum A region of space that appears to be empty (a vacuum) but actually contains stored energy. When this stored energy is released, the false vacuum is said to decay. (See vacuum.)

field theory A theory in which forces are communicated between two particles by an energy "field," which fills up the space between the two particles. In a field theory, any particle, such as an electron, is surrounded by a field. The field continuously creates and destroys intermediary particles, which transmit the force of the electron to other particles. In fact, particles themselves are considered to be concentrations of energy in the field.

flat universe A universe that is at the boundary between an open and closed universe. In a flat universe, the average mass density always has precisely the critical value necessary to keep the gravitational energy equal to the energy of expansion. Therefore the value of omega is 1 for that universe. Flat universes also have infinite size and the geometry of an infinite, flat surface, that is, Euclidean geometry. (See closed universe; critical mass density; Euclidean geometry; omega; open universe.)

flatness problem The puzzle of why the universe today is so close to the boundary between open and closed, that is, why it is almost flat. Equivalently, why should the average mass density today be so close to the critical mass density, but not exactly equal to it? If omega begins bigger than 1, it should get bigger and bigger as time goes on; if it begins smaller than 1, it should get smaller and smaller. For omega to be near 0.1 today, about 10 billion years after the big bang, it had to be extraordinarily close to 1 when the universe was a second old. Some people consider such a fine balance to have been highly unlikely according to the standard big bang model and thus are puzzled as to why the universe today is almost flat. (See closed universe; critical mass density; flat universe; omega; open universe.)

fluctuations Deviations from uniform conditions. For example, a mass of gas that bunches to a higher density than the surrounding gas would be referred to as a fluctuation. The amount of bunching for each scale of mass is called the fluctuation spectrum. Most cosmologists attempt to explain the observed structures in the universe (such as groups of galaxies) by the gravitational condensation and growth of small fluctuations of mass in the past.

Friedmann equation An equation for the evolution of the universe. The Friedmann equation can be derived from Einstein's theory of gravity, plus the assumptions that the universe is homogeneous (looks the same at every point) and isotropic (looks the same in every direction). The solution of the Friedmann equation tells, among other things, how the distance between galaxies changes with time. (See homogeneity; isotropy.)

Friedmann models A general class of cosmological models that assume the universe is homogeneous and isotropic on large scales and that allow the universe to evolve in time. Most calculations in the standard big bang model assume a Friedmann cosmology. (See Friedmann equation; homogeneity; isotropy.)

fundamental constants of nature Physical quantities, like the speed of light or the mass of an electron, that enter into the laws of physics in a basic way and are believed to be the same at all times and everywhere in the universe. Most physicists take the fundamental constants as given properties of the universe.

galaxy An isolated aggregation of stars and gas, held together by their mutual gravity. A typical galaxy has about 100 billion stars, has a total mass equal to about a trillion times the mass of the sun, is about 100,000 light years in diameter, and is separated from the nearest galaxy by a distance of about a hundred times its own diameter, or 10 million light years. Thus, galaxies are islands of stars in space. Our galaxy is called the Milky Way. Galaxies come in two major shapes: flattened disks with a central bulge, called spirals, and amorphous, semispherical blobs, called ellipticals. If galaxies are found bunched up next to one another, they are said to lie in groups or clusters. Clusters with a particularly large number of galaxies in them are called rich clusters. Galaxies that do not lie in such groups, but rather seem to be scattered uniformly and randomly through space, are called field galaxies. Some galaxies are characterized by the dominant type of radiation they emit. For example, radio galaxies are unusually strong emitters of radio waves.

general relativity Einstein's theory of gravity, formulated in 1915. The theory of general relativity prescribes the gravity produced by any arrangement of matter and energy.

globular cluster A spherical congregation of stars, within a galaxy, that orbit one another because of their mutual gravity. A typical globular cluster has about a million stars; thus globular clusters are much smaller than galaxies. There are about 100 globular clusters in our galaxy, the Milky Way.

grand unified theories (GUTs) Theories in physics that attempt to explain the forces of nature as manifestations of a single underlying force.

gravitational constant A fundamental constant of nature that measures the strength of the gravitational force. The gravitational constant is also called Newton's gravitational constant and is denoted by G.

gravitational force The weakest of the four fundamental forces of nature, the gravitational force between any two masses is proportional to the product of the masses and varies inversely as the square of the distance between them. The other three fundamental forces are the electromagnetic force and two kinds of nuclear forces. (See electromagnetic force; nuclear force.)

Great Attractor A hypothesized large mass, some 100 million light years from earth, that seems to be affecting the motions of many nearby galaxies by virtue of its gravity.

GUTs See grand unified theories.

hierarchical clustering model A model of galaxy clustering in which different patterns appear at different scales of distance, indefinitely to larger and larger scales, and in which the "average" density of matter depends on the size of the volume over which the average is performed. In a homogeneous model, on the other hand, the average density is independent of the size of the volume over which the average is performed. (See pancake model.)

homogeneity In cosmology, the property that any large volume of the universe looks the same as any other large volume. Most cosmological models assume homogeneity.

horizon The maximum distance that an observer can see. In cosmology, the horizon is equal to the distance that light has traveled since the beginning of the universe. Objects more distant than our horizon are invisible to us because there has not been enough time for light to have traveled from there to here.

horizon problem The puzzle that widely separated regions of the universe are observed to share the same physical properties, such as temperature, even though these regions were too far apart when they emitted their radiation to have exchanged heat and homogenized during the time since the beginning of the universe. In particular, we detect the same intensity of cosmic radio waves (cosmic background radiation) from all directions of space, suggesting that the regions that emitted that radiation had the same temperature at the time of emission. However, at the time of emission, when the universe was about 300,000 years old, those regions were separated by roughly 50 million light years, much exceeding the distance light or heat could have traveled since the big bang. (See horizon.)

Hubble constant The rate of expansion of the universe. The Hubble constant actually changes in time, even though it is called a constant, because gravity is slowing down the rate of expansion of the universe. The Hubble constant is equal to the recessional speed of a distant galaxy, divided by its distance from us. If we assume a homogeneous and isotropic universe, the recessional speed of a distant galaxy is proportional to its distance; thus the Hubble constant as determined by any receding galaxy should be the same, yielding a universal rate of expansion of the universe. According to estimates, the current value of the Hubble constant is approximately 1 per 10 billion years, meaning that the distance between any two distant galaxies will double in about 10 billion years at the current rate of expansion. Astronomers measure the Hubble constant in units of kilometers per second per megaparsec. For example, a Hubble constant of 100 kilometers per second per megaparsec—which astronomers would refer to simply as a Hubble constant of 100—corresponds to 1 per 10 billion years. The Hubble constant is denoted by the symbol H_0. (See Hubble law.)

Hubble law The law that recessional speed is proportional to distance for a homogeneous and isotropic universe. Galaxies moving away from us with a speed precisely following this law are said to follow the Hubble flow. Because the actual universe is not precisely homogeneous, with lumpiness arising from clustering of galaxies and voids of empty space, the motions of actual galaxies deviate somewhat from the Hubble flow.

Hubble time The reciprocal of the Hubble constant. The Hubble time (or period) gives an estimate for the age of the universe. To obtain a precise value for the age of the universe, omega must also be known, since the expansion rate has changed in time. (See deceleration parameter; Hubble constant; omega.)

hydrodynamics The study of how gases and fluids flow under applied forces.

image tubes Electronic devices that amplify incoming light while preserving its direction.

inflationary universe model A recent modification of the standard big bang model in which the infant universe went through a brief period of extremely rapid (exponential) expansion, after which it settled back into the more leisurely rate of expansion of the standard model. The period of rapid expansion began and ended when the universe was still much less than a second old, yet it provides a physical explanation for the flatness and horizon puzzles. The inflationary universe model also suggests that the universe is vastly larger than the portion of it that is visible to us. (See exponential expansion.)

initial conditions (See boundary conditions.)

isotropy In cosmology, the property that the universe appears the same in all directions. The uniformity of the cosmic background radiation, coming from all directions of space, suggests that on the large scale the universe is isotropic about our position. If we then assume that our position is not unique, we conclude that the universe appears isotropic about all points. This last result requires that the universe be homogeneous. (See cosmic background radiation; homogeneity.)

kiloparsec A measure of distance equal to 1,000 parsecs, or about 3,000 light years.

large-scale structure The distribution of galaxies and other forms of mass on large distance scales, covering hundreds of millions of light years and larger. A perfectly homogeneous and isotropic universe would have no large-scale structure; a universe with all the galaxies lined up in single file would have enormous large-scale structure.

large-scale motions Bulk motions of distant galaxies deviating from the Hubble flow. (See Hubble law.)

leptons Fundamental particles of nature, which may interact via all of the fundamental forces except the strong nuclear force. The electron is an example of a lepton.

luminosity The energy per second emitted by an astronomical object, analogous to the wattage of a light bulb.

Mach's principle The hypothesis that the inertia of bodies, that is, their resistance to acceleration by applied forces, is determined not by any absolute properties of space but by the effects of distant matter in the universe. Equivalently, Mach's principle proposes that the distinction between accelerated and nonaccelerated frames of reference is determined by the effects of distant matter.

many-worlds (Everett-Wheeler) interpretation of quantum mechanics The view of quantum mechanics holding that a physical system simultaneously exists in all of its possible states, prior to and after a measurement of the system. (Compare with the Copenhagen interpretation of quantum mechanics.) In the many worlds interpretation, each of these simultaneous existences is part of a separate universe. Every time we make a measurement of a physical system and find it to be in a particular one of its possible states, our universe branches off to one of the universes in which the system is in that particular state at that moment. The system, however, continues to exist in its other possible states, in parallel universes. (See Copenhagen interpretation of quantum mechanics; quantum mechanics.)

mass-to-light ratio (M/L) The ratio of total mass in a physical system to the amount of radiation produced by that system. Often, mass that does not produce radiation of any kind can nevertheless be detected by its gravitational effects. Systems with a large amount of such dark matter would have a high mass-to-light ratio.

matter-to-antimatter ratio The ratio of mass in particles to mass in antiparticles. For every type of particle, there is an antiparticle counterpart. The positron, for example, is the antiparticle of the electron and is identical to the electron except for having opposite electrical charge. The abundances of particles and antiparticles do not have to be equal. It appears that our universe is made up almost entirely of particles, rather than antiparticles, although there is no fundamental difference between the two kinds of matter.

megaparsec A measure of distance equal to 1 million parsecs, or about 3 million light years.

metals In astronomy, all elements heavier than hydrogen and helium, the two lightest elements.

minute of arc A unit of angle equal to 1/60 of a degree.

missing mass The cosmic mass that some scientists hypothesize so that the universe will have the critical density of matter, with an exact balance between gravitational energy and kinetic energy of expansion giving omega = 1. Such mass is called missing because it represents about 10 times as much mass as has actually been detected by any method, including studies of gravitational effects. (See closed universe; critical mass density; dark matter.)

mixmaster model A non-Friedmannian cosmological model that begins with a highly anisotropic infant universe and shows how anisotropies are reduced in time. (See Friedmann models.)

N-body simulations Computer simulations of the behavior of a large number of bodies under their mutual interactions. In cosmological N-body simulations, the bodies are usually galaxies and the interactions are gravitational. Thus, the computer simulates how a group of galaxies should behave under their mutual gravitational attraction. The law of gravity and the initial positions and velocities of the hypothetical galaxies and other masses are fed into the computer. The computer then calculates the evolution of the system.

neutrino A subatomic particle that has no electrical charge, has little if any mass, and interacts with other particles only through the weak nuclear force and the gravitational force. (See nuclear forces.)

neutron A type of subatomic particle that, together with the proton, makes up the atomic nucleus. The neutron has no electrical charge and is composed of three quarks. (See proton; quark.)

new inflation A 1982 modification of the original inflationary universe model. While the original inflationary universe model solved a number of cosmological problems, it led to the result that the universe was very inhomogeneous during the inflationary epoch and contained bubbles of empty space surrounded by a medium filled with energy. In new inflation, no such bubbles appear, although the universe still undergoes a brief epoch of extremely rapid expansion. (See inflationary universe model.)

no-boundary proposal An initial (boundary) condition for the universe proposed by Stephen Hawking and James Hartle. In this proposal, the mathematics of general relativity are reformulated so that time is replaced by a space-like coordinate, in effect representing the universe as having 4 space dimensions instead of 3 space dimensions and a time dimension. (In such a formulation, "time" does not have its usual meaning.) Hawking and Hartle suggest that the geometry of this representation of the universe should be analogous to the geometry of the surface of a sphere, that is, a shape with no edges—hence the name no-boundary proposal. When translated back into ordinary time and space, this suggested boundary condition takes the form of a specific initial condition for the universe. The no-boundary proposal is formulated within a quantum mechanical calculation of the behavior of the early universe, although such calculations are incomplete because of the lack of a self-consistent theory of quantum gravity.

non-Euclidean geometry Geometry that does not follow the postulates and results of Euclidean geometry. For example, in a non-Euclidean geometry, the sum of the interior angles of a triangle differs from 180 degrees. According to Einstein's general relativity theory, gravity distorts space into a non-Euclidean geometry.

nuclear forces There are two kinds of nuclear forces: the strong nuclear force and the weak nuclear force. These two forces, plus the gravitational and electromagnetic forces, comprise the four fundamental forces of nature. The strong nuclear force, which is the strongest of all four forces, is the force that holds protons and neutrons together in the atomic nucleus. The weak nuclear force is responsible for certain kinds of radioactivity; for example, the disintegration of a neutron into a proton, electron, and antineutrino.

nucleon A proton or neutron. (See neutron; proton.)

nucleosynthesis The production of heavy nuclei from the fusion of lighter ones. According to the big bang theory, the infant universe consisted of only hydrogen, the lightest of all atomic nuclei, because any heavier nuclei would have come apart in the intense heat. All other elements would have to be formed later, in nucleosynthesis processes. It is believed that most of the helium, the next lightest element after hydrogen, was formed when the universe was a few minutes old, and that most of the other elements were made much later, in nuclear reactions at the centers of stars.

old inflation The original (1981) inflationary universe model. (See inflationary universe model; new inflation.)

omega The ratio of the average density of mass of the universe to the critical mass density, the latter being the density of mass needed to eventually halt the outward expansion of the universe. In an open universe, omega is always less than 1; in a closed universe, it is always greater than 1; in a flat universe it is always exactly equal to 1. Unless omega is exactly equal to 1, it changes in time, constantly decreasing in an open universe and constantly increasing in a closed universe. Omega has been measured to be about 0.1, although such measurements are difficult and uncertain. (See closed universe; critical mass density; flat universe; open universe.)

open universe A universe fated to expand forever. In an open universe, the kinetic energy of expansion is always greater than the gravitational energy, and the value of omega is always less than 1. Open universes have the geometry of an infinite curved surface with the same amount of curvature at every point. (See closed universe; omega.)

order-of-magnitude estimate An approximate estimate of the magnitude of something, accurate to within a range of 10 times too big to 10 times too small. For example, given that the population of the United States is 250 million, any estimate of the population lying between 25 million and 2,500 million would be an acceptable order-of-magnitude estimate. Astronomers are accustomed to order of magnitude estimates.

oscillating universe (model) A universe that expands, then contracts, then expands, then contracts, and so on through many cycles.

pancake model A model of galaxy formation in which the first structures to condense out of the smooth background of primordial gas were very large in size. These large masses then collapsed into thin sheets (pancakes) and fragmented into many smaller pieces the size of galaxies. A competing theory, sometimes called the hierarchical clustering model, proposes that the first structures to form were the size of galaxies. As galaxies clustered together, due to gravity, larger and larger structures were formed. (See hierarchical clustering model.)

particle physics That branch of physics that attempts to understand the fundamental particles and forces of nature.

particle-to-antiparticle ratio Same as matter-to-antimatter ratio.

Perseus–Pisces region A region of space containing a huge congregation of galaxies called a supercluster. The galaxies in this supercluster appear to be distributed in a long chain.

peculiar velocity A deviation in the velocity of a galaxy from that expected on the basis of a uniform expansion of the universe. (See Hubble law.)

photon The subatomic particle that transmits the electromagnetic force. Light consists of a stream of photons.

photon-to-baryon ratio The ratio of the number of photons to the number of baryons in any typical, large volume of space. (See baryons; photon.)

Planck's constant A fundamental constant of nature that measures the magnitude of quantum mechanical effects. Visible light, for example, consists of discrete particles of light, or photons, each carrying an amount of energy equal to Planck's constant multiplied by the frequency of visible light. (The energy of one photon of visible light is approximately 10^{-18}, or a billionth of a billionth, the energy of a penny dropped to the floor from waist high.) By combining Planck's constant with two other fundamental constants of nature—Newton's gravitational constant and the speed of light—one obtains other Planck units that mark critical densities and times when quantum mechanics and gravity were both extremely important. For example, the Planck density, or Planck scale, is the density of matter above which the structure, and perhaps meaning, of space and time break down due to quantum mechanical effects. Numerically, the Planck density is about 10^{93} grams per cubic centimeter. The infant universe had this enormous density when it was about 10^{-43} seconds old, which is called the Planck time, and when it had a temperature of about 10^{22} Centigrade. At this temperature, the mean energy per particle was equivalent to the Planck mass, about 10^{-5} grams. (See quantum mechanics.)

Planck mass (See Planck's constant.)

Planck time (See Planck's constant.)

population I, II, and III stars The youngest observed stars are called population I stars; older observed stars are called population II; and it is postulated that an even older generation of stars, called population III, existed still earlier. Population II stars formed mostly from hydrogen and helium. Population I stars, like our sun, formed from hydrogen, helium, and a large range of heavier elements (like carbon and oxygen) believed to have been created in the interiors of earlier population II and III stars and then blown out into space.

proton A type of subatomic particle that, together with the neutron, makes up the atomic nucleus. The proton has a positive electrical charge and is made up of three quarks. (See neutron; quark.)

q_0 The deceleration parameter. (See deceleration parameter.)

quantum cosmology The subfield of cosmology that deals with the universe during its first 10^{-43} seconds, when quantum mechanical effects and gravity were both extremely important. (See Planck's constant; quantum mechanics.)

quantum field A distribution of energy that is constantly creating and destroying particles, according to the probabilities of quantum mechanics, and transmitting the forces of nature. (See field theory; quantum mechanics.)

quantum fluctuations Continuous variations in the properties of a physical system, caused by the probabilistic character of nature as dictated by quantum mechanics. For example, the number of photons in a box with perfectly reflecting walls constantly varies because of quantum fluctuations. Quantum fluctuations can cause particles to appear and disappear. Some theories hold that the entire universe was created out of nothing, in a quantum fluctuation.

quantum gravity A theory of gravity that would properly include quantum mechanics. To date, there is no complete and self-consistent theory of quantum gravity, although successful quantum theories have been found for all the forces of nature except gravity. (See quantum mechanics.)

quantum mechanics The theory that explains the dual wave-like and particle-like behavior of matter and the probabilistic character of nature. According to quantum mechanics, it is impossible to have complete and certain information about the state of a physical system, just as a wave cannot be localized to a single point in space but spreads out over many points. This uncertainty is an intrinsic aspect of the system or particle, not a reflection of our inaccuracy of measurement. Consequently, physical systems must be described in terms of probabilities. For example, in a large collection of uranium atoms, it is possible to accurately predict what *fraction* of the atoms will radioactively disintegrate over the next hour, but it is impossible to predict *which* atoms will do so. As another example, an electron with a well-known speed cannot be localized to a small region of space but behaves as if it occupied many different places at the same time. Any physical system, such as an atom, may be viewed as existing as a combination of its possible states, each of which has a certain probability. Quantum theory has been extremely successful at explaining the behavior of nature at the subatomic level, although many of its results violate our commonsense intuition. (See Copenhagen interpretation of quantum mechanics; many-worlds interpretation of quantum mechanics; uncertainty principle; wave function.)

quark One of the fundamental, indestructible particles of nature, out of which many other subatomic particles are made. The neutron, for example, is built of 3 quarks. Five types of quarks have been discovered, and it is believed that a sixth also exists. Quarks interact mainly via the strong nuclear force and the electromagnetic force.

quasars Extremely distant and luminous astronomical objects that are much smaller than a galaxy and much more luminous. Quasars may be the central regions of certain very energetic galaxies, at an early stage of their evolution. It is believed that the power of a quasar derives from a massive black hole at its center.

redshift A shift in color toward the red end of the spectrum, caused when a source of light (and color) is moving away from the observer. The magnitude of the redshift is directly related to the magnitude of recessional speed; thus measurement of the redshift of an object measures its recessional speed. In an expanding universe, the colors of galaxies are shifted to the red, and in a uniformly expanding universe, the redshift is directly proportional to an object's distance from earth (except for extremely distant objects). Measurement of an astronomical object's redshift provides the distance to the object.

redshift survey The methodical tabulation of the redshifts of a large number of galaxies in a particular region of the sky. Redshifts directly measure the recessional speed of galaxies. If Hubble's law is assumed, these speeds can be translated to distances. Under such an assumption, a redshift survey provides the third dimension, depth, for the galaxies in a survey. The other two dimensions for each galaxy are provided by its perceived position on the sky. The redshift of a galaxy is obtained by measuring its spectrum of light; in this way it is possible to see how much its colors are shifted. (See spectrum.)

Riemannian geometry A large class of non-Euclidean geometries. The mathematics of general relativity uses Riemannian geometry. (See general relativity.)

relativity The theory of how motion and gravity affect the properties of time and space. The special theory of relativity establishes, among other things, the nonabsolute nature of time. The amount of time elapsed between two events will not be the same for two observers or clocks in relative motion to each other. The general theory of relativity describes how gravity affects the geometry of space and the rate at which time passes. (See general relativity; special relativity.)

reproducing universes The process in some inflationary universe models whereby the universe is constantly spawning new universes, causally disconnected from one another and from the parent universe. (See chaotic inflation.)

Robertson-Walker metric A mathematical description of the geometric properties of a homogeneous and isotropic universe. Friedmann cosmologies all use the Robertson-Walker metric. (See Friedmann models; homogeneity; isotropy.)

scale factor A measure of distance in cosmology. The distance between any two galaxies, for example, is proportional to the scale factor, which is always increasing in an expanding universe. If the scale factor doubles in size, then the distance between any two galaxies doubles.

scale length A measure of the size of a physical system or region of space.

Schrödinger's equation A fundamental equation in quantum mechanics for the evolution of the wave function of a system. (See quantum mechanics; wave function.)

Schwarzschild radius or limit The "surface" of a black hole, within which the strength of gravity is so strong that light cannot escape. The Schwarzschild radius is proportional to the mass of the black hole and would be about 2 miles for a black hole with a mass equal to that of our sun. Black holes were first "theoretically discovered" by Karl Schwarzschild in 1917. (See black hole.)

Schwarzschild singularity The center of a black hole. According to Einstein's theory of general relativity, the entire mass of a black hole is concentrated at a point at its center, the "singularity." Physicists today believe that quantum mechanical effects, not included in the theory, would cause the mass to spread out over a tiny but nonzero region, thus preventing an infinite density of matter and doing away with the singularity.

second law of thermodynamics A physical law formulated in the nineteenth century and stating that any isolated system becomes more disordered in time. (See entropy.)

simulations In science, models of the behavior of physical systems made with a computer. (See N-body simulations.)

singularity A localized place, either in space or in time, at which some quantity, such as density, becomes infinite. The laws of physics cannot describe infinite quantities and, in fact, physicists believe that infinities do not exist in nature. All singularities, such as the Schwarzschild singularity, are therefore probably the artifacts of inadequate theories rather than real properties of nature. According to Einstein's theory of general relativity, the universe began in a singularity of infinite density, the big bang. Physicists now believe that an improved and yet-to-be discovered modification of general relativity, incorporating quantum mechanics, will show that the universe did not begin as a singularity. (See Schwarzschild singularity.)

singularity theorems In astronomy and cosmology, mathematical proofs that show the conditions under which a mass will gravitationally collapse to form a singularity. The singularity theorems of cosmology, proved in the 1960s, indicate that the current behavior of the universe, together with the laws of general relativity but without quantum mechanical corrections, require that at some definite time in the past the universe was compressed to a state of zero size and infinite density, called a singularity. The laws of physics break down at a singularity and cannot be used to predict anything during or before the singularity occurred. (See singularity.)

special relativity Einstein's theory of time and space. Special relativity, formulated in 1905, shows how measurements of length and time differ for observers in relative motion.

spectrograph An instrument that records the amount of light in each range of wavelength, that is, in each range of color. In general, each type of astronomical object, such as a star or a galaxy, will emit a characteristic spectrum of light. (See spectrum.)

spectroscopy The study of what wavelengths of light an object or substance will emit under various conditions.

spectrum The amount of light in each range of wavelength, that is, in each range of color. The term spectrum can also be applied more generally to the intensity of something at each length scale. An object that emits radiation in a continuous range of colors is said to have a continuous spectrum. An object that emits radiation only at certain wavelengths is said to have emission lines; an object that *absorbs* radiation only at certain wavelengths is said to have absorption lines.

standard candle In astronomy, any class of objects with the same luminosity, or with some property that allows a reliable determination of the luminosities. (See Cepheid; luminosity.)

statistical distribution The range of variation of some quantity in a population, obtained by sampling many members of the population. For example, the statistical distribution of the height of American males could be obtained by sampling 10,000 randomly chosen males and counting the number of them within each range of heights. In cosmology, the distance between pairs of galaxies, averaged over a large number of galaxies, would constitute a statistical distribution.

steady state model A model of the universe in which the universe does not change in time. A modern steady state model was proposed in the late 1940s. The big bang model is not a steady state model.

super cluster A cluster of galaxy clusters, extending about 100 million light years.

thermal equilibrium The condition of a system in which all its parts have exchanged heat and have come to the same temperature. An isolated system in thermal equilibrium does not change over time. This is also a state of maximum disorder. (See black body radiation; entropy.)

thermodynamics The study of how bodies change as they exert forces and exchange heat with other bodies.

Tully–Fisher relation An observed relation between the luminosity of a spiral galaxy and the rotational speed of its stars. More luminous galaxies have stars that are moving faster. (See galaxy.)

uncertainty principle Also called the Heisenberg uncertainty principle, a fundamental result of quantum mechanics stating that the position and speed of a particle cannot be simultaneously known with complete certainty. If one is known with high certainty, the other becomes very uncertain. The product of uncertainty in position and uncertainty in momentum (mass multiplied by speed) is equal to a constant, Planck's constant. Since both initial position and initial speed are required to forecast the future position of a particle, the uncertainty principle prevents completely accurate predictions of the future from the past, even if all the laws of physics are known. (See Planck's constant; quantum mechanics.)

vacuum A state of minimum energy. Empty space is often referred to as the vacuum. Because of the uncertainty principle, even empty space has a minimum energy content.

velocity field The velocities of a group of objects with different velocities at different positions of space.

virial theorem (or method) In gravitational physics, a quantitative relationship between the amount of gravitational energy and the amount of kinetic (motional) energy of an isolated physical system in equilibrium. Thus, for such a system, only one of the two kinds of energy need be directly measured; the other can be inferred by use of the virial theorem. The universe as a whole is not in a state of equilibrium; its gravitational energy and kinetic energy of expansion are thus not required to obey the virial theorem.

voids Large regions of space without galaxies.

wave function The mathematical description of a physical system according to the laws of quantum mechanics. The wave function tells what possible states the physical system could be in and what is the probability of being in any particular state at any given moment.

weak interactions Interactions between certain particles caused by the weak nuclear force between them. (See nuclear force.)

Weinberg–Salam model A theory in physics, developed by Steven Weinberg, Abdus Salam, and Sheldon Glashow, that unifies two fundamental forces of nature, the electromagnetic force and the weak nuclear force. This is the same as the electroweak theory.

Acknowledgments

This little book grew out of the introduction to *Origins: The Lives and Worlds of Modern Cosmologists*, a previous book by myself and Roberta Brawer concerned with the personal styles of working cosmologists. I am grateful to Roberta for her contributions to that introduction.

For advice on scientific matters I thank Edmund Bertschinger, Margaret Geller, Paul Schechter, and Edwin Turner, with special appreciation to Martin Rees for his comments on several versions of the manuscript. For other critical comments on the evolving manuscript, I thank Susan Carey, Owen Gingerich, James Peebles, Andrew Pickering, Martin Rees, Carol Rigolot, Paul Schechter, Diane Steiner, Edwin Turner, and Charles Weiner, with special gratitude for the help of Richard Goodwin and Michelle Preston. My editors at Harvard University Press, Angela von der Lippe and Susan Wallace, made many fine suggestions about content, presentation, and editing of the manuscript. Laszlo Meszoly contributed to the book's clarity with his excellent line drawings. Finally, I wish to thank Jane Gelfman for her continued advice and support in literary matters.

Illustration Credits

Unless otherwise indicated, all line drawings are by Laszlo Meszoly.

CHAPTER ONE

Assyrian cylinder, courtesy of Bibliothèque Nationale, Paris.

"Expulsion of Adam and Eve from Paradise," courtesy of Robert Lehman Collection, The Metropolitan Museum of Art (1975.1.31).

Copernican world system, courtesy of the Royal Astronomical Society.

Thomas Digges' world system, from Francis R. Johnson, *Astronomical Thought in Renaissance England* (Baltimore: Johns Hopkins University Press).

CHAPTER THREE

Title page of Einstein's paper from *Sitzungsberichte der Preussiche Akademie der Wissenschaften* (1917).

Albert Einstein, photograph by John Hagemeyer, Bancroft Library, courtesy of the American Institute of Physics Niels Bohr Library.

Alexander Friedmann, courtesy of the Leningrad Physico-Technical Institute and the American Institute of Physics Niels Bohr Library.

Georges Lemaître, courtesy of Owen Gingerich.

Henrietta Leavitt, courtesy of Harvard College Observatory.

Andromeda galaxy, courtesy of Harvard College Observatory.

Edwin Hubble, courtesy of the Observatories of the Carnegie Institution of Washington.

Sombrero galaxy, courtesy of National Optical Astronomy Observatories.

Hubble's diagram from *Proceedings of the National Academy of Sciences* (1929).

CHAPTER FOUR

Allan Sandage, photograph by Douglas Cunningham, courtesy of the Observatories of the Carnegie Institution of Washington.

Fred Hoyle, photograph by Lotte Meitner-Graf.

James Peebles, photograph by Robert P. Mathews, courtesy of Princeton University.

Robert Dicke, photograph by Robert P. Mathews, courtesy of Princeton University.

Arno Penzias and Robert Wilson, courtesy of AT&T Archives.

COBE, courtesy of NASA Goddard Space Flight Center.

CHAPTER FIVE

Maarten Schmidt, courtesy of California Institute of Technology.

Martin Rees, courtesy of Martin Rees.

CHAPTER SIX

Charles Misner, photograph by Cindy Grimm, courtesy of the Campus Photo Services, University of Maryland, College Park.

Roger Penrose, courtesy of Roger Penrose.

CHAPTER SEVEN

Gérard de Vaucouleurs, courtesy of University of Texas.

Margaret Geller, photograph by Steven Seron, courtesy of the Center for Astrophysics.

John Huchra, photograph by Steven Seron, courtesy of the Center for Astrophysics.

1986 Center for Astrophysics redshift survey, courtesy of V. de Lapparent, M. J. Geller, and J. P. Huchra, *Astrophysical Journal Letters* 302 (1986): L1.

Great wall, courtesy of M. J. Geller and J. P. Huchra, *Science* 246 (1989): 897.

Vera Rubin, courtesy of Robert J. Rubin.

Alan Dressler, courtesy of Carnegie Institution of Washington.

Sandra Faber, photograph by Don Fukuda, courtesy of the University of California, Santa Cruz.

Jeremiah Ostriker, courtesy of Jeremiah Ostriker and Princeton University Department of Physics.

Computer simulations, courtesy of James M. Gelb and Edmund Bertschinger of MIT.

Survey of 2 million galaxies, courtesy of S. J. Maddox, G. P. Efstathiou, W. Sutherland, and J. Loveday, Oxford University, Department of Astrophysics.

CHAPTER EIGHT

Mt. Palomar Telescope, courtesy of Palomar Observatory.

Keck telescope, courtesy of *Sky and Telescope*.

Hubble Space Telescope, courtesy of Lockheed and NASA.

Z-machine, courtesy of Dave Penland, Smithsonian Institution.

CCD, courtesy of G. Luppino.

James Gunn, photograph by Robert P. Mathews, courtesy of Princeton University.

Infrared Milky Way, courtesy of NASA Goddard Space Flight Center/COBE Science Working Group.

AXAF, courtesy of TRW and NASA.

SIRTF, courtesy of NASA.

CHAPTER NINE

Stephen Hawking, photograph by Clyde S. White, Carolina Biological Supply Co.

CHAPTER TEN

Steven Weinberg, courtesy of Steven Weinberg.

Alan Guth, photograph by Alan Lightman.

Andrei Linde, courtesy of Andrei Linde.

Index

Designed by Marianne Perlak

Typeset in Linotron Bembo

Library of Congress Cataloging-in-Publication Data

Lightman, Alan P., 1948–
Ancient light : our changing view of the universe / Alan Lightman.
p. cm.
"Adapted from Origins: the lives and worlds of modern cosmologists, by Alan Lightman and Roberta Brawer, published by Harvard University Press, 1990"—T.p. verso.
Includes bibliographical references (p.) and index.
ISBN 0-674-03362-0
1. Cosmology. 2. Astronomers. I. Lightman, Alan P., 1948– Origins. II. Title.
QB981.L538 1991
523.1—dc20

91-12459
CIP

COLLEGE OF MARIN
3 2555 00075446 0